The Institute of Biology's
Studies in Biology no. 5

Plant Taxonomy

by V. H. Heywood, *Ph.D. (Cantab.), D.Sc. (Edin.)*
Professor of Botany, The University of Reading

Edward Arnold (Publishers) Ltd

First published 1967
Reprinted 1968
Reprinted 1970

Printed in Great Britain by
William Clowes and Sons Ltd, London and Beccles

General Preface to the Series

It is no longer possible for one textbook to cover the whole field of Biology and to remain sufficiently up to date. At the same time students at school, and indeed those in their first year at universities, must be contemporary in their biological outlook and know where the most important developments are taking place.

The Biological Education Committee, set up jointly by the Royal Society and the Institute of Biology, is sponsoring, therefore, the production of a series of booklets dealing with limited biological topics in which recent progress has been most rapid and important.

A feature of the series is that the booklets indicate as clearly as possible the methods that have been employed in elucidating the problems with which they deal. There are suggestions for practical work for the student which should form a sound scientific basis for his understanding.

1967

INSTITUTE OF BIOLOGY
41 Queen's Gate
London, S.W.7.

Preface

After a period of quiescence taxonomy has again come to occupy a major, some would say central, role in biology. In the days when there is a strong tendency towards the unification of biology into a physico-molecular framework, taxonomy has been described as the lonely voice speaking on behalf of an interest in diversity. It also serves as a focus by which the diverse ramifications of biology can be seen in perspective. Taxonomy also extends beyond the frontiers of biology and, as Lincoln Constance has recently said, there is surely no aspect of biology more deeply intertwined in man's history, economy, literature, aesthetics and folklore, with its overt concern for the variety of organisms in the world.

Recent developments in taxonomy have been so numerous and far-reaching that only an outline can be given in this booklet but I have attempted to weave them into a coherent account of the field as a whole.

V. H. H.

Liverpool
August, 1966.

Contents

The Diversity of Nature and the Need for Classification 1

Faced with the vast array of diversity in the natural world, man instinctively classifies. He divides this diversity up into smaller, more manageable groups. Even in the early stages of civilization man soon came to recognize which kinds of plants he could eat, which he could use for fuel, which would poison him. These early groupings were practical and determined largely by their economic importance. They were not based entirely on morphology, although the appearance of the plants was used for recognition, but also on smell, taste, nutritive value. Later certain conspicuous structural features of plants impressed themselves on man so that at an early stage groups such as the Umbelliferae with their characteristic inflorescence, Cruciferae with their typical flowers and Leguminosae with their characteristic flowers and fruits, were recognized and described.

Classification is the basic method which man employs to come to grips with and organize the external world. Plants and animals are in fact classified in basically the same way as non-living objects—on the basis of possession of various characters or relations which they have in common. Which kinds of features are employed depends on the kind of classification we have in mind, since classification is done for a purpose and different classifications are needed for different purposes. Just as we can classify library books by the colour of their bindings or by their authors, so we can classify plants by the colour of their flowers (as in handbooks for amateurs, permitting easy identification) or by their pollination type. It is easy to see that in each of these examples the usefulness of the classification varies according to one's viewpoint. They are, however, of little general value, and the most commonly used biological classifications are based on large amounts of data and are designed to serve more purposes.

We shall consider the different kinds of classification later, but it is important to recognize that some adequate method is necessary to allow us to identify parts of the organic world and to communicate such information to other people. Classification is an information storage and retrieval system: without it there would be chaos.

The following estimates give us some idea of the size of the problem (Table 1).

Table 1 Estimated numbers of described species of plants (after GRANT, 1963).

Flowering plants	286,000	Fungi and slime moulds	40,400
Gymnosperms	640	Protista	30,000
Ferns and fern allies	10,000	Blue-green algae	1,400
Bryophytes	23,000	Bacteria	1,630
Algae (green, red, brown)	8,675	Viruses	200

Biological classification is required today for several reasons. Biologists and others need to have a reference system for the plants they work with. They work with named entities—named usually to the extent of species, genus and, by implication, family. The reason for requiring a name is two-fold: it provides a reference point, but at the same time behind the name lies a whole series of assumptions—that there is a group of individuals sharing certain features in common so that whenever a plant is determined as belonging to a named species there is a high degree of probability that it will always have the same sort of characteristics.

Classification today tends to be taken for granted. It forms a built-in system of guarantees that biologists and other scientists enjoy and assume to be effective. The very act of naming gives species or other groups an apparent reality which may bear no relation to the biological facts. The name as such is not important—any other name or reference device such as a number would suffice (there are many situations where numbers are more useful than names in systematics): it is the biological information behind the name that is important to the users of taxonomy. When a physiologist or ecologist or horticulturist identifies, or has determined for him, a plant as belonging to, say, *Poa annua*, he does not question what this means in terms of classification. He accepts that, apart from errors and minor variations, plants referred to as *P. annua* from any part of the world where they grow will be essentially similar to each other, not only in morphology (otherwise they would not be recognizable) but in their general characteristics of structure, physiology, biochemistry, etc.

This important characteristic of the biological classification we use today is called *predictivity* or *predictive value*. It does not happen by chance, but is built in by the very methods used by systematists to make their classifications. By predictivity is meant that if a class of plants shares many characteristics in common, when some feature not used in constructing the class is found in one member, then the other members will probably also possess it. Although systematists achieve a high degree of predictivity in their classifications by basing them on multiple correlations of characters, this does not of course explain why such correlations are possible, or in fact why classification of living things in a generally acceptable way is possible at all.

There is a strong factor at work affecting the features of both plants and animals. It is, of course, their evolutionary history, as has been realized since 1859 when DARWIN published the *Origin of Species*. An important distinction must be made here between the fact that a classification is *made possible* by evolutionary history or phylogeny and actually *basing* the classification on this history or phylogeny. The evolutionary relationships of plants are highly complex and cannot be reduced to a single factor on which classification could be based. The relationship between evolution and classification is of great importance in systematics and is discussed in Chapter 4.

Obviously evolution introduces a new dimension in systematics since another important aspect of classification is relationship. Evolutionary relationships are stressed by many taxonomists but there are various other kinds of relationship which may or may not be expressed by a classification: overall relationship (or affinity), which is derived from the resemblances and differences shown by the organisms, genetic relationship, chemical relationship, and so on.

The term *systematics* is used to cover the scientific study of the diversity and differentiation of organisms and the relationships which exist between them. *Taxonomy* is that part of systematics which deals with the study of classification, including its bases, principles, procedures and rules. Often the terms are used as interchangeable. *Classification* in a biological sense is the process of ordering plants into groups which are arranged hierarchically; the term is also used for the ensuing arrangement. A taxonomic group of any rank (species, genus, etc.) is called a *taxon* (pl. taxa).

1.1 The major phases of systematics

In the pre-evolutionary period most of the data available to taxonomists were morphological and anatomical, coupled with increasing knowledge of geographical distribution. This morpho-geographical approach reached its culmination towards the end of the nineteenth century and is still an important component of present-day taxonomy.

The effect of evolutionary theory on taxonomy and in particular on classification was remarkably small in practice although it provided a radically new conceptual framework. Phylogenetic considerations entered into taxonomy and the reasons for the similarity between present-day organisms became understandable as being the consequence of common descent and common genetic makeup. The relationship between classification and evolution has been a major source of debate since the beginning of the evolutionary period.

What is seldom acknowledged is that Darwinian evolution provided no new techniques and no new procedures for taxonomists to employ in constructing classifications. As a consequence there was no great advance in the classifications produced in the post-Darwinian period that could be attributed to evolutionary ideas. What did result was a powerful reason for interpreting classifications arrived at by conventional means in evolutionary terms, but there are grounds for believing that speculative attempts to produce phylogenetic trees and systems without much valid evidence actually retarded the progress of taxonomy. We did, however, learn a great deal about the operation of natural selection and the ways in which plant populations behave in nature.

This new understanding of the dynamic nature of populations and their biological makeup developed in the next major phase of taxonomy—the *cytogenetical-biosystematic* period which began in the 1920's and continued to dominate the scene till about 1960. In this period rapid developments in cytology and genetics were applied to taxonomy, and chromosome data, particularly karyotype studies, polyploidy, crossability and breeding behaviour (see Chapter 10) replaced to a large extent conventional taxonomic approaches, especially in the universities. The New Systematics of the 1940's was really little more than an acknowledgement of the importance of these approaches to taxonomy and was marked by the publication of a number of major text-books collecting together and synthesizing the results.

The effects on classification were mainly at the level of the species and population: two of the most important were the introduction and acceptance of the 'biological species' concept which sought to define species in terms of gene pools and breeding barriers, and the recognition of the value of chromosome numbers as good 'marker characters' for the delimitation of taxonomic groups and for working out the evolutionary sequence of groups.

New data from anatomy, pollen, embryology, etc., were applied to classifications above the level of species, but few procedures for the handling of these data and the actual construction of classifications were produced. The evolutionary interpretation of classifications, however, continued apace.

We are now entering a new period marked by major developments of techniques and ideas in the whole field of systematics. This period has been referred to as a taxonomic revolution or explosion. The two main areas of advance are in the fields called Chemosystematics and Taxometrics or Numerical Taxonomy. The contribution from chemosystematics is two-fold: (*a*) it provides a new class of data for use in constructing or modifying classifications; (*b*) it gives us a valuable means of assessing the probabilities of certain kinds of phylogenetic relationships, particularly those referring to common ancestry and evolutionary sequences. Very recent developments suggest that biochemical data, especially at the 'macromolecular' level, may assist in working out some of the evolutionary pathways by which groups have arisen (see p. 38).

Taxometrics, on the other hand, is a field primarily concerned with procedural or operational problems, that is to say, with the actual steps we have to follow to make classifications using the data provided from various sources, and more recently with attempts at reconstructing evolutionary relationships using numerical means. It is not concerned with producing new data but with methods of handling them by means of electronic computers so as to reduce the subjective element involved in comparing sets of data (see p. 44).

We can summarize the present situation in taxonomy by saying that, in addition to a consolidation of our knowledge of the variation and differentiation of populations and of the mechanisms causing it, there is an increasing tendency to use biochemical data which more directly reflect the genetic makeup of organisms, to quantify and assess data numerically whenever possible, and to reduce the subjective or intuitive elements involved in classification so that we can see not only the results but the ways by which they have been achieved step by step.

A unique feature of taxonomy is that none of its sources of data ever becomes obsolete. We cannot dispense with morphology and anatomy just because we have cytological and chemical data. There is, on the contrary, a succession of new classes of data, each consolidating what has gone before. For this reason we shall consider not only the latest developments in taxonomy but also the present state of the conventional approaches.

Although taxonomy was the earliest of the biological sciences, it has not yet completed its basic task of surveying and classifying organic diversity, to say nothing of understanding its structure and evolution in detail. It is a moot question whether we will be able to discover and describe all the plant species in unknown tropical regions before the areas are destroyed by advancing civilization. Even in Europe, the cradle of scientific taxonomy, several hundred species of flowering plants and ferns were described as new in the period 1945–60.

One of the major needs today is a cataloguing of the world's genetic resources so that they can be conserved and utilized in breeding programmes for future generations. The richest sources of genetic variability are in gene centres such as the Mediterranean and the Near East where there is a serious lack of floristic knowledge, that is, of the identity and distribution of the plants that occur there, and Floras of these regions are urgently required.

The Raw Material and the Basic Groups—Populations and Species 2

It is an observable fact of nature that plants and animals occur in large numbers—in populations—and that at the same time they appear to be grouped together as different 'kinds'—what we call species. Populations and species in their different ways form the basis of classification.

2.1 Populations

The term population is widely used by biologists but with different meanings. Loosely speaking it refers to those plants or animals (or both) that grow in a particular habitat at a particular time. So the plants growing in a field or in a pond or on a mountain-top form local populations. In this sense several populations may go to make up a species. In the wider sense populations can be regarded as all the plants throughout the distribution range that can be regarded as belonging to a given species. In such cases there are bound to be discontinuities between the different parts of the population, for geographical reasons if no other, and we are really dealing with a system of populations. The more widespread such population systems are, the more likely there are to be appreciable differences between them. The common foxglove, *Digitalis purpurea*, is widespread in West Europe, occurring frequently in the British Isles in local stands; the British plants belong to the typical subspecies *purpurea*, while in various parts of Spain, Portugal, Corsica and the Balearic Islands, other subspecies replace it. Even within the range of the populations of subspecies *purpurea* there are marked differences between those north and south of the Pyrenees—differences in habit, seed colour, type of hairs, etc. On a smaller scale there are differences between the members of the British foxglove populations in respect of markings of the corolla and whether the stems are hairy or glabrous. The proportion of hairy-stemmed individuals (caused by a single gene) varies in different parts of the British Isles and appears to be higher in the north.

When populations of foxglove are found in the field it is a good exercise to score, say, 50 or 100 plants for glabrous or hairy stems. The results for different populations can then be compared to see if any geographical pattern emerges.

As might be expected, plants of the same species growing together will tend to interbreed if they are genetically intercompatible, and the term population is often used in this restricted sense, as a breeding group, by some biosystematists and geneticists. Attention is focussed on the local breeding populations—those plants in a given area that can exchange genes—since they are of considerable evolutionary importance. This idea can be extended further by considering populations as consisting of all those plants of a species that actually, or potentially, exchange genes. This is a largely theoretical concept since it is concerned not only with what actually happens at the present day but what could happen if plants, today widely separated geographically and not in breeding contact, were to meet and cross, either directly or through intermediates. From a genetical or evolutionary viewpoint such a view of populations is clearly of great importance. Some taxonomists feel that species should be defined in such a way as to coincide with such theoretical populations, but it is not the kind of definition that can be put into practice, however attractive in theory (see p. 56).

2.2 Local breeding population

The importance of the local breeding population, whose members grow together and can interbreed to form a common gene pool, has been generally recognized by biosystematists and geneticists for it is the unit of evolutionary change. Clearly, it is also of great importance for taxonomy since our classifications deal with the way plants are organized in nature.

Populations whose members reproduce sexually are called *amphimictic* and, of these, those between whose members there is potentially free gene exchange are called *panmictic*. Such panmictic populations show variation in many features: no two are alike in composition or in habitat conditions or in size, even though they belong to the same species. The fact that they are recognizable as separate populations means that there is some degree of isolation between them. In general the more isolated from each other populations are, the more likelihood there is that they will continue to evolve in respect of several of their features and become more distinct. The rate of change depends on the size of the populations, mutation rate, the breeding system, the habitat features and the pressure of natural selection.

Populations then are obviously dynamic and we have to study the variation between the individuals *within* the populations and the variation *between* populations. The variation between individuals is caused by three factors: (1) modifications due to the external environment; (2) mutation; (3) genetic recombination. The actual pattern that the variation takes is largely determined by the breeding system and the whole process is under the control of natural selection. These are considered in later sections.

2.3 Breeding system

Whether the different members of a population can interbreed depends to a large extent on the breeding system which operates. The breeding system is one of the main factors which regulate the amount of genetic recombination in plants. From both a taxonomic and an evolutionary viewpoint the system of recombination is of great importance since it regulates the actual production of variability in a population by restricting the kinds of gametes formed and the types of zygotes produced.

Although we may talk about populations as gene pools (all the individuals contributing to the pool) the genes do not float around freely in the pools as individuals allowing unrestricted recombination. In fact the genes are grouped in chromosomes and in breeding groups so that there is canalization of their movement at the moments of meiosis and fertilization. Thus the amount of recombination actually achieved differs from one kind of organism to another. Some systems of recombination are closed as in *apomictic* plants, which do not reproduce sexually (see p. 10), where the only sources of variation are mutation and other special methods; others are relatively restricted where there are serious barriers to recombination such as partial or predominant self-fertilization; others are open where there are no main barriers to recombination and a large number of different kinds of genotypes can be formed. The main types of factors regulating recombination are:

1. Length of generation
2. Chromosome number
3. Frequency of crossing over
4. Breeding system
5. Pollen dispersal
6. Diaspore dispersal
7. Population size
8. Isolating mechanisms (genetic and external)

Recombination can easily be shown by sowing a packet of *Phlox drummondii*, 'Twinkle', and selecting, say, four different types of flower colour patterns from the array of plants that come up. If plants of these four types are then selfed it will be found that a large number of recombinant types will be found in their progeny and difficulty will be experienced in selecting pure lines.

The *breeding system* refers to the various physiological and morphological mechanisms which control the relative frequency of cross or self fertilization in a population or taxonomic group.

Many different mechanisms have been evolved by plants to promote or restrict cross fertilization, ranging from self-incompatibility and separation of the sexes (dioecism) thus ensuring outcrossing, to failure of the flowers to open (cleistogamy) enforcing self fertilization. Every transition between these extreme conditions is found and the commonest condition is for there to be a mixed economy of crossing and selfing in a population (see Table 2).

Table 2 Breeding systems.

I. Sexual (Amphimixis)
  (*a*) Inbreeding (autogamy)
    e.g. cleistogamy
  (*b*) Outbreeding (allogamy)
    1. Mechanical features of flower preventing self-pollination
    2. Temporal features having a similar effect
      e.g. protandry, protogyny
    3. Monoecism
    4. Gynodioecism or androdioecism
    5. Differential growth of pollen tubes
    6. Genetic incompatibility
    7. Heterostyly
    8. Dioecism
II. Asexual (Apomixis)
  (*a*) Vegetative reproduction—with or without sexual reproduction
  (*b*) Agamospermy—sporadic, facultative, obligate

This can be understood if we consider the way in which the different breeding systems adapt populations to different environmental conditions. Outcrossing populations store a large amount of variability in recessive genes in the heterozygous condition. This variability is available for future use by recombination, but since most of it is non-adaptive it reduces the immediate fitness of the population under more or less constant environmental conditions. If conditions change there is a good possibility that some of the genotypes being continually released by recombination will be adapted to the new conditions. Inbreeders, on the other hand, tend towards the homozygous condition and store up variability *between* different homozygous lines into which the population is divided. This variability can be released by occasional outcrossing. Inbreeders possess immediate fitness and are often found colonizing new, relatively uniform habitats; their lack of flexibility, however, makes them incapable of adjusting to changing conditions.

Populations usually show a balance between these two situations of immediate fitness and long-term fixity (inbreeders) and reduced immediate fitness and potential flexibility (outbreeders).

The breeding system can be tested quite simply in some species. For example, if a single plant of *Anthoxanthum odoratum* is grown in a pot and covered with nylon bolting cloth, no seed will be set. If, however, several plants are grown together and put under one cover, there will be a good seed set. This is because the plants are self-incompatible (due to a genetic mechanism) and a single plant cannot therefore fertilize itself and produce seed, whereas if several are grown together the effect of the wind is sufficient to cause pollen transfer and consequent cross fertilization leading to seed set. Other self-incompatible species that can be tested include *Ranunculus flammula*, *Trifolium repens*, *Achillea millefolium*, *Calystegia sepium*, *Holcus lanatus*, *Lotus corniculatus* and *Euphorbia cyparissias*.

Care should be taken not to assume the breeding system from the structural features of the flower, since many species showing apparent attraction devices favouring cross pollination, such as showy petals, nectaries, etc. are apomictic. A classic example is *Taraxacum officinale*, the dandelion, which has a complex stylar brush mechanism favouring outcrossing but is in fact apomictic although the stimulus of pollination is needed for the productior of apomictic seed.

2.4 Non-sexual populations: apomixis

In many groups of plants sexual reproduction has been replaced or substituted by non-sexual methods. The term *apomixis* is used for this type of breeding system, and it includes *vegetative reproduction*, where what are normally accessory means of reproduction such as bulbils, tubercles, stolons, offsets, etc. take over the whole reproductive role, as well as *agamospermy* where seeds and embryos are produced by non-sexual processes.

The populations built up by vegetative reproduction are called *clones* and extensively used in agriculture and horticulture, for example potatoes, dahlias, etc. Several clones found in nature are of hybrid nature such as *Acorus calamus*, which is represented in Europe by a sterile triploid which spreads vegetatively; other examples of clonal hybrids are found in the genus *Mentha*.

In some cases sexual taxa may show a high rate of vegetative reproduction giving a similar effect to vegetative apomixis as in *Calystegia pulchra*, an Asiatic species which has spread extensively in the British Isles and in other parts of Europe by the reproduction of a small number of clones introduced last century.

Bulbils may replace some or all of the flowers in the inflorescence as in some species of *Allium*, grasses, especially Arctic and Alpine ones as in *Poa*, *Festuca*, *Deschampsia*, and this phenomenon is known as *pseudo-vivipary*. The degree of pseudo-vivipary may be under environmental control and can differ from country to country, as in the case of *Festuca vivipara* which is regarded as probably having been derived from *F. ovina* which is non-viviparous. Agamospermy is known in more than 40 families of Angiosperms and has been widely studied in *Taraxacum*, *Alchemilla*, *Rubus*, *Sorbus*, etc. The offspring of agamospermous plants are genetically identical with their parent and form a clone, and taxonomists have, in the past, been tempted to describe each of these clones as a separate 'species' since they are constant in their characteristics and show little variation, even though they are separable from one another by only very slight character differences. Consequently hundreds of apomictic 'species' have been published in genera such as *Rubus*, *Taraxacum* and *Ranunculus* (the *auricomus* group).

Agamospermy occurs in varying degrees in different groups, ranging from a low frequency in several species through to a nearly obligate condition in other taxa. Partial (facultative) agamospermy in a group results in the formation of complex patterns of variation: the sexually reproducing plants will produce a new series of biotypes (plants with identical or similar genetic makeup) which will segregate and may become frozen again as a set of apomictic clones. Even in species where apomixis is apparently obligate, some degree of sexual reproduction probably occurs. As in the relationship between inbreeding and outbreeding in species, it appears that in some groups of predominantly apomictic species such as *Poa pratensis* and *P. alpina* a balance is struck between sexually produced and agamospermous seed so that hybridization between the apomictic biotypes followed by recombination will allow for the production of new biotypes in one or two generations. Hybridization is often closely linked with the evolution of apomictic groups and there is also a close association with polyploidy (the occurrence of multiple chromosome sets—see p. 52). The consequent variation patterns are frequently so complex that no formal taxonomic treatment is possible.

2.5 The problem of inbreeders

The populations of habitually inbreeding plants normally consist of one or a few homozygous biotypes, causing variation *between* the populations to be discontinuous. This results in the formation of numerous morphologically uniform groups, as in apomictic taxa (although for a different reason) which can be treated as distinct taxa and easily separated from one another. Many of the scores of microspecies described by JORDAN in genera such as *Erophila* were of this nature. Again like apomicts, inbreeding occurs in different degrees and the completely obligate condition probably does not occur in flowering plants. In *Thlaspi alpestre*, for example, it is estimated that 5 per cent outbreeding may occur and such outcrossing between inbreeding lines allows the production of hybrids which will be available for selection to produce new inbreeding lines. The result of a mixed economy of selfing and crossing in facultative inbreeding populations is the formation of complex variation patterns which again, as in the case of facultative apomicts, may be difficult to classify satisfactorily. In both cases it is as much the associated hybridization as the breeding mechanism which causes taxonomic problems, and although we may understand the causes of the variation, this does not tell us how to handle the results taxonomically.

The Taxonomic Structure 3

3.1 The taxonomic hierarchy: groups and categories

Biological classification is built up from the raw material of populations and species according to a set of simple principles. In the last chapter we considered the species as groups of individuals, sampled from populations, closely resembling each other in most characters. Those species which share most characters in common are placed together into larger (more inclusive) groups called *genera*; these in turn are assembled into yet more inclusive groups called *families*, and so on. This system of building up a series of increasingly inclusive groups on the basis of overall similarities is sometimes referred to as the box-within-box method which is a simplified version of set theory (Fig. 3–1), although it should be noted that the relative size of the boxes varies from group to group.

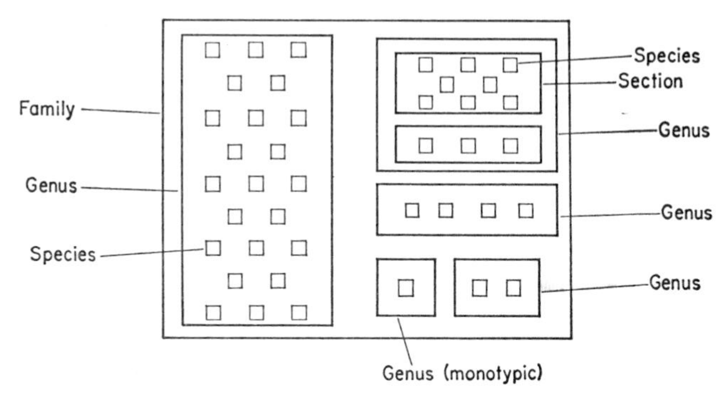

Fig. 3–1 Showing the box-within-box arrangement of taxonomic groups.

Parallel with these different levels of groups is a series of categories or levels forming a hierarchy. The hierarchy of categories is like a set of empty shelves arranged at different heights; once formed the groups have to be assigned to a particular category. It is the category into which we place a group that determines its group name. The categories in the hierarchy are called species, genus, family, class, etc. (Fig. 3–2) and the groups placed at these levels are given the same name. Deciding upon the rank to be given to a group is often a difficult task in classification since there is no way of defining what is meant by any given rank or category except by saying what its position is relative to those above and below it.

It has been arbitrarily decided over the course of the past two centuries how many levels or categories are necessary to accommodate the variation of the whole plant kingdom. The actual levels of distinctness of the groups placed at these levels are decided upon by comparison with other groups and, to a considerable extent, by tradition. In the case of species and family there tends to be less difficulty in making a decision. This is because, in

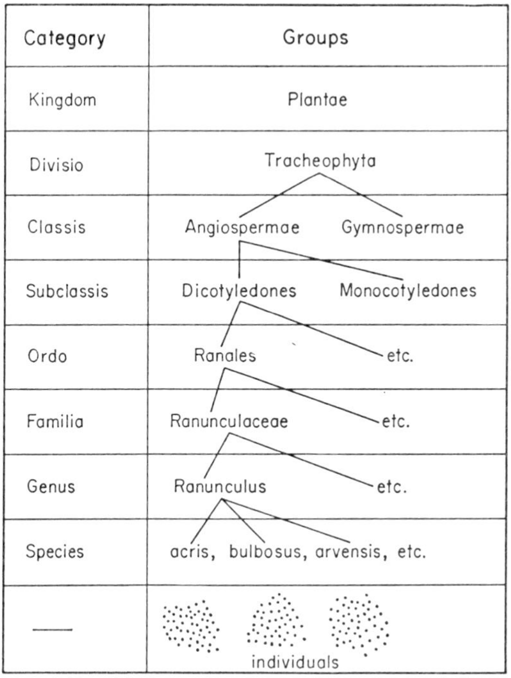

Category	Groups
Kingdom	Plantae
Divisio	Tracheophyta
Classis	Angiospermae Gymnospermae
Subclassis	Dicotyledones Monocotyledones
Ordo	Ranales etc.
Familia	Ranunculaceae etc.
Genus	Ranunculus etc.
Species	acris, bulbosus, arvensis, etc.
—	individuals

Fig. 3–2 Showing the way individuals are placed into a group (species) which in turn is a member of successively more inclusive groups, and the corresponding categories in which these groups are accommodated.

theory at least, species are non-arbitrary as to both inclusion and exclusion (Fig. 3–3): this means that there is continuous variation between all the members of a species and complete discontinuity of variation in some features between one species and another. As we have indicated, in defining species we look for complete discontinuities in a number of correlated features. Families are recognizable because they represent major peaks of evolutionary variation and, since there are relatively few of them, taxonomists seldom have to concern themselves with their delimitation.

Between the family and the species there is no easy way of deciding on the appropriate degree of inclusiveness and there is no way of defining a genus, for example, in such a way that is not equally applicable to a subgenus or subfamily or tribe. For this reason what is regarded as a single genus by one taxonomist may be regarded as a tribe containing several genera by another taxonomist, without any change in the characters of the plants concerned. Recently attempts have been made to quantify the levels

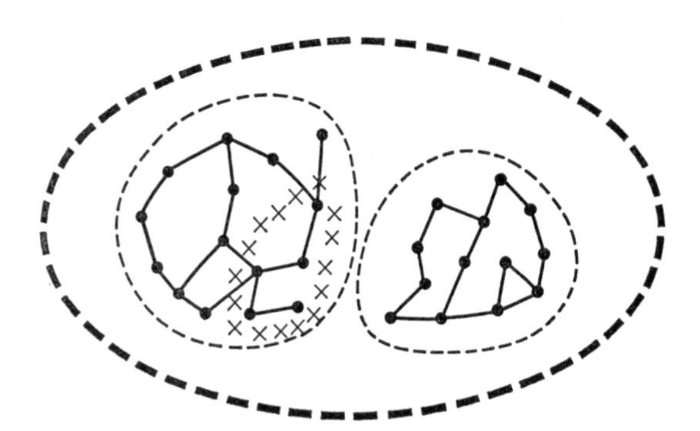

Fig. 3–3 Arbitrary and non-arbitrary nature of taxa. Species are indicated by light broken lines, containing individuals (heavy dots) which are linked to indicate their continuity of variation. Each is non-arbitrary as to inclusion in the sense that all its members are continuous in variation, and non-arbitrary as to exclusion in that it is discontinuous from other groups in variation. The two species are included in a heavy broken line, indicating a genus which is arbitrary as to inclusion since it has internal discontinuities, but non-arbitrary as to exclusion. In the left-hand species, a subspecies is indicated by crosses. It is non-arbitrary as to inclusion, but arbitrary as to exclusion as it has an external continuity. (After SIMPSON, 1961, *Principles of Animal Taxonomy*, Columbia Univ. Press, New York; Oxford Univ. Press London).

of difference for the different ranks so that they can be applied more or less uniformly within particular families or parts of families. Fig. 3–4 shows how this may be done by assessing the overall similarity between species in respect of characters measured. The level still has to be chosen subjectively and will vary from family to family. A possible solution may arise from molecular biology (see Chapter 7).

3.2 Names or numbers

This flexibility of the hierarchical system of classification can prove a liability since the genus name is both part of the classification and part of the name of the organism. This has two main disadvantages: (1) it forces us to deal with a cumbersome code of rules which regulate the nomenclature of plants; (2) if we decide that a species, for example *Ranunculus ficaria*, is better placed in another genus (*Ficaria*) on the basis of its overall similarities, then its generic name (and in this case its specific name, too) has to be changed (becoming *F. verna*). As a reference system, therefore, it is clearly defective.

To meet this situation proposals have been made to institute a system of uninomial nomenclature in which present-day generic and specific names would be united or hyphenated to form uninomials which would be stabilized, governed by simple rules and freed from many of the restrictions of a nomenclatural code. Genera and higher taxa would be relatively informal groupings. For example, instead of having the alternative binomials

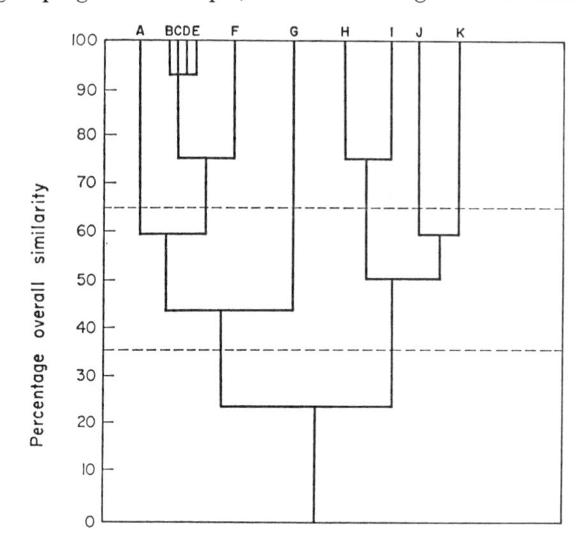

Fig. 3–4 Dendrogram indicating overall resemblance between eleven imaginary species. BCD and E show 95 per cent similarity in the features measured and could be regarded as components of a species aggregate. H and I show a 75 per cent similarity while both of them are only 45 per cent similar to J and K. The horizontal broken line drawn arbitrarily at 35 per cent similarity divides the species into two groups which might be regarded as subgenera or sections. The horizontal broken line at 65 per cent similarity divides the species into six groups which could be regarded as series. If 'standard' levels of similarity are employed to decide upon different ranks within the same family, a more objective use of categories will be possible.

Ranunculus ficaria and *Ficaria verna*, the Lesser Celandine would always be referred to by the uninomial *Ranunculus-ficaria* (or *R.-ficaria*) regardless of whether the species was considered as belonging to *Ranunculus* or *Ficaria*, or to any other genus for that matter.

A further step is to introduce a system of numbering to replace the names. There would be two numbers for each taxonomic group—an arbitrary *reference number* which would normally remain unchanged for a species no matter how a classification changed, and a *classification* number which would explain by means of its digits the classificatory position of an organism. The *International Plant Index* which is currently being developed is designed to provide all plants with such a classification number. In this index the family Ranunculaceae has the number 1331.7794 in which 1 = Dicotyledones, 1331 = Ranales; the family Mucoraceae is 7108.6123 in which 7 = Fungi, 71 = Phycomycetes, 7108 = Mucorales.

An additional advantage of a numbering system is that it would be suitable for automated information storage and retrieval machinery which is being increasingly used in biological classification. Clearly a numbering system does not have the mnemonic value of a naming system. The present binomial system was introduced largely to allow biologists to remember the names of organisms and at the same time to indicate their systematic position. Linnaeus expected all well-educated botanists to know *all* the genera! With well-known plants these two advantages still hold—*Ranunculus ficaria* or *Pinus sylvestris* are meaningful to British botanists, but there are many thousands of generic names and the vast majority of them are unfamiliar and therefore convey little to most biologists. The probable solution lies in having a parallel system of names and numbers for different purposes.

The Two Sides of the Coin 4

The kinds of classification we have considered so far are *empirical*, i.e. they are based on greater or lesser numbers of characters which can be observed and assessed; the groupings are formed (at least in theory) almost mechanically by placing together those individuals or groups of them which share the largest numbers of characters, following logical principles. They are the conventional classifications which serve as an information retrieval and storage system for biologists. The characters employed are those of present-day organisms and no evolutionary principles are involved, although the organisms themselves are of course the end points of evolutionary history.

The term *phenetic* is also applied to these classifications to distinguish them from phylogenetic (or phyletic) ones. Any sources of data, other than phylogenetic ones, may be used in the construction of phenetic classifications, so that chemical, cytological, embryological and other features may be included as well as the traditional sources of morphology and anatomy, but there is no implication as to the evolutionary significance of the groups so produced.

The similarities shown by the members of phenetic groups are probably the result of common evolutionary ancestry and this was usually assumed to be so by post-Darwinian taxonomists, thus allowing phenetic groups such as the various Angiosperm families to be regarded as phyletic groups. This need not necessarily be true, and in fact it is seldom possible to demonstrate how true it is in any particular cases. The reason for this difficulty is that

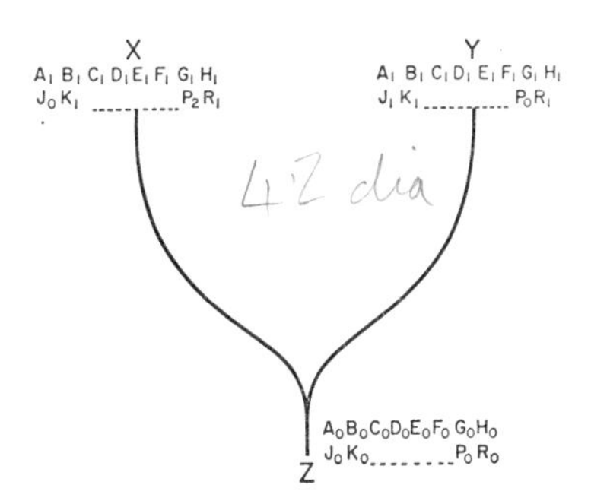

Fig. 4–1 Primitive patristic similarity between lineages *X* and *Y* and the common ancestor *Z* from which they have diverged. Characters A–H have not changed in the descendants from the state (indicated by the subscript 0) in which they occur in the ancestor. Evolutionary changes in the other characters (J, K . . . P, R) differ in the two descendants. (After SOKAL and CAMIN, 1965, *Systematic Zoology*, **14**, 184.)

not all evolution is divergent; there are in fact several different but inter-related evolutionary processes we must consider if we are to make phylogenetic interpretation of a classification.

Resemblance in phenetic groups due to common ancestry is termed *patristic*. This in turn can be divided into (1) *primitive patristic similarity* (Fig. 4–1) where two or more evolutionary lines have been derived from a common ancestor and retain the same characters as the common ancestor apart from those derived from initial splitting into two or more lines; thus both the ancestor and the descendants share with each other the same number of unchanged characters; and (2) *derived patristic similarity* where the characters in the descendants (apart from those which have changed by the initial splitting) are not the same as in the common ancestor because they have become independently modified in the same way (Fig. 4–2). In the

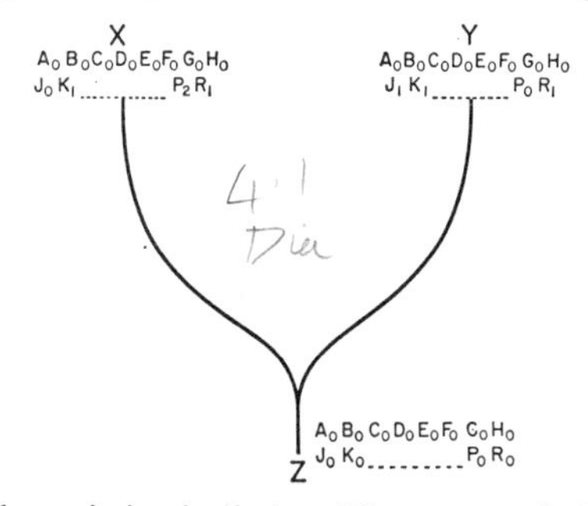

Fig. 4–2 Derived patristic similarity. The states of characters A–H in the descendants *X* and *Y* are not identical with those of the ancestor *Z* (as in Fig. 4–1) since the characters have become independently modified to state 1. *X* and *Y* show the same phenetic similarity as in Fig. 4–1, but the similarity of either with *Z* is much less. The element of similarity between *X* and *Y* is derived patristic similarity. (After SOKAL and CAMIN, 1965, *Systematic Zoology*, **14**, 184.)

latter case the degree of similarity between the descendants is the same as in the previous example due to parallel evolution, but the similarity of either with the common ancestor is much reduced. It should be noted that unless we knew the features of the common ancestor we could not distinguish between the two kinds of patristic affinity in the derived groups.

Phenetic similarity may also be due to convergence, when it is termed *homoplastic*. The term *homoplasy* is defined as resemblance not due to inheritance from common ancestry and includes convergence and parallelism, two evolutionary phenomena which are difficult to define in a practical way. *Parallelism* (Fig. 4–3(b)) can be defined in a general sense as the development of similar features separately in two or more, genetically similar, fairly closely related lineages; if it is restricted to cases where the lineages are of common ancestry and the similarity is developed on the basis of the characteristics of that ancestry then it can be included under derived patristic similarity (above). *Convergence* (Fig. 4–3(a)) is the development of similar features separately in two or more genetically diverse and not closely related lineages and not due to a common ancestry. In this sense the

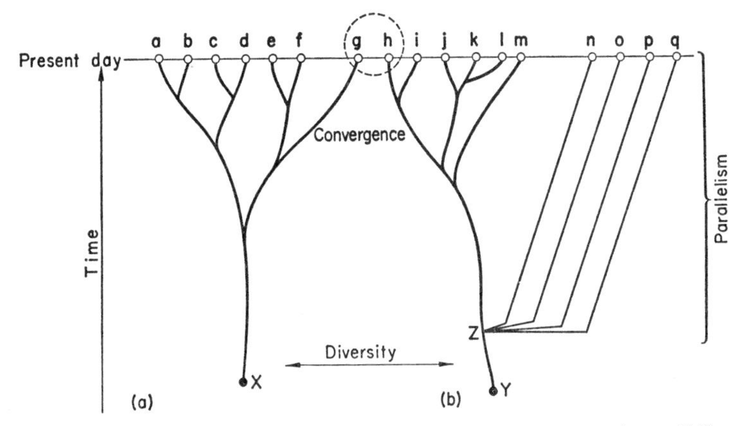

Fig. 4–3 (**a**) Convergence, shown by *g* and *h* which come from different phyletic lines, *X* and *Y*. (**b**) Parallelism, shown by *n*, *o*, *p* and *q* which have a common origin at *Z* and have subsequently evolved in parallel.

common acquisition by closely related species of buttercups of the aquatic habit and finely dissected leaves is parallelism, and the largely superficial resemblances between some fleshy cacti, senecios and euphorbias is conveniently called convergence.

In practice there are all intermediate situations between the two and indeed the terms are often interchanged in the literature. When we further consider that we seldom have the necessary evolutionary and genetic information available to distinguish between the two situations or even decide if either is operative, involving decisions as to whether characters are homologous or analogous, we see how exceedingly difficult it can be to tell what the causes of phenetic resemblances are.

Even if we are able to decide in a particular case that the cause of resemblance between present-day groups is due to patristic similarity, this tells us nothing of actual pathways by which it has been derived nor when the changes took place.

Cladistics is that part of phylogenetic relationship that refers to the pathways of evolution. If we represent it in the form of a tree or dendrogram it is the study of how many branches there are and which branch arose from which and in what sequence. The difference between cladistic and patristic relationship is shown in Fig. 4–4. There are difficulties in estimating the degree of cladistic affinity between taxa in such a diagram: we can use the number of divergences that one has to pass before one taxon is united to another or we can take into account the actual time of origin of

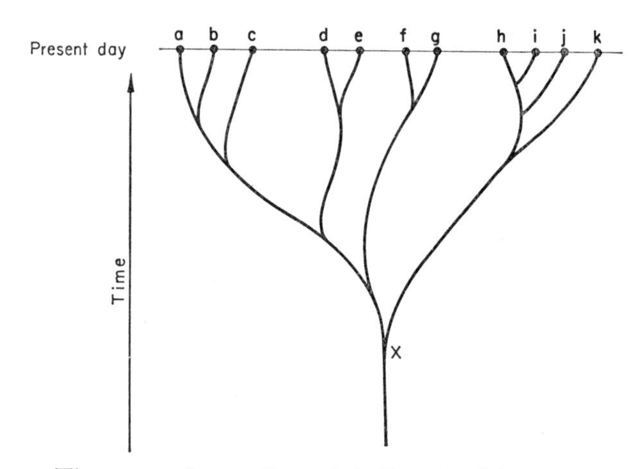

Fig. 4–4 The taxa *a–k* are all *patristically* related in that they have a common ancestry through the stem at *X*. Each taxon has a different degree of *cladistic* relationship with each other. The pairs *a* and *b*, *d* and *e*, *f* and *g*, *h* and *i*, are separated by only one divergence before they retrace common ancestry. Between *a* and *k* there are five divergences after they leave the common stock at *X*.

one taxon relative to another, and there are other methods of assessing the ways in which any taxon in the diagram is related to any other.

The time element is an important component of phylogenetic relationships. This is called *chronistics* and can be defined as the study of the time scale during which evolutionary events occur. In many phylogenetic diagrams chronistic relationship is indicated by the vertical axis: sometimes it is explicitly indicated, sometimes it is implied, often it is ignored. Even in patristic relationships there is clearly a distinction to be made between simultaneous acquisition of derived similarity in Fig. 4–2 in both lineages and a different timing of the evolutionary steps involved in each line.

Having considered the theoretical background to phylogenetic relationships let us now consider this in relation to actual classifications. Although it is the aim of the majority of taxonomists to produce evolutionary classifications, it must be obvious that the concept of evolutionary or phylogenetic classification covers a wide range of different approaches. A fully phylogenetic classification must express not only phenetic relationships but also include and distinguish between patristic, cladistic and chronistic dimensions. So far no means has been devised of depicting a phylogenetic tree within the framework of a hierarchical classification, which is what is apparently aimed at.

In practice most taxonomists are satisfied with a phenetic classification in which the groups are (*a*) monophyletic, i.e. had a common ancestry, (*b*) *arranged* relative to one another in a scheme or tree which indicates their approximate evolutionary relationship. The various components discussed above are seldom even recognized, largely because the evidence is not available so the question of applying it does not arise!

What are the sources of evidence? For common ancestry, fossil evidence would be desirable, but since it is seldom available in bacteria, fungi, mosses or even ferns and flowering plants, the only approach possible is to attempt to produce phenetic groups based on overall resemblances and then attempt to remove those features of similarity that are due to convergence. There is no certain method of achieving this, but in some cases there may be strong evidence from chromosome number, biochemical features and certain morphological characters to indicate which member of a group is really an outsider. Examples are given in Chapters 6 and 7.

Fossil evidence again is normally necessary to assess the details of cladistic relationship. In its absence we must again turn to our present-day phenetic groups and from various sources of evidence work out the probabilities of the sequence in which they evolved. This method cannot provide detailed pathways of branching since we do not know how many lineages have arisen and died out in the past during the evolution of our contemporaneous groups. The most fruitful approach is to consider the evolutionary sequences of individual characters or organs—*semophyleses*, as opposed to whole taxa. *Primitive* characters are those features of taxa which were possessed by their ancestors; *advanced* characters are those that show a greater or lesser degree of specialization in proportion to their departure from the ancestral condition.

The different organs of a plant do not evolve at the same rate. Thus the genus *Delphinium* in the Ranunculaceae is clearly specialized in its zygomorphic spurred perianth, yet primitive in possessing follicles. This makes it difficult to assess the overall degree of specialization of a plant. In the absence of fossil evidence it has been necessary to postulate a large number of dicta about the primitive and specialized condition of various characters of angiosperms. Those suggested by BESSEY and HUTCHINSON are largely accepted by taxonomists today: for example, trees and shrubs are usually more primitive than herbs; spiral arrangement of leaves and perianth parts is more primitive than opposite and whorled; flowers with many parts are less advanced than those with few parts; hypogyny is the primitive condition, perigyny and epigyny being derived from it, etc. In addition to these trends in morphological characters and others like them, there are many more in features of pollen morphology and anatomy, wood and floral anatomy, embryology, the karyotype, and from the biochemistry of secondary plant products.

The latter two sources, karyology and biochemistry, are particularly useful since the evidence is often clear cut, unambiguous and, what is more important, the *direction* of the trends is usually indisputable. This is important since many morphological trends are clear enough but it is difficult to be sure of the direction since trends can go into reverse in cases where selection pressures favour it. Moreover they may be read from any point in the sequence, in either direction.

The evolution of organs is quite different from the evolution of taxa, and one can work out the overall degree of specialization of a taxon by considering as many characters as possible. Such an advancement index is, however, an imaginary average condition but it is often the most that can be done by way of presenting estimates of evolutionary status against a phenetic background. Examples of this approach are illustrated by SPORNE (1956) in a review of the subject with reference to the families of the Dicotyledones.

Methods for reconstruction of cladistic relationships are currently being developed employing statistical techniques and numerically produced phenetic classifications (Chapter 9), but are still very primitive and depend largely on assumptions about the evolutionary trends in the characters employed.

It must be emphasized that a full understanding of the systematics of plants will only come from a study of all three relationships—phenetic, cladistic and chronistic. Classifications can be and are produced by phenetic methods alone and serve a vital function by organizing the facts we obtain about nature; they also provide a framework for studies about the processes and mechanisms of evolution. Cladistic relationships alone will not provide a classification of any general value. Unfortunately no methods are yet available to produce classifications combining all these elements and some would claim that they never will be. What is evident is that the better our phenetic groupings are, the better our chances of evolutionary interpretation will be.

Phenotype, Genotype and Natural Selection 5

5.1 Phenotype and genotype

The taxonomist works not only with the outer appearance of organisms (the *phenotype*) but with their genetic makeup (the *genotype*). In the last chapter we considered the changes in the heredity and characteristics of populations in the course of evolution. Now we have to look at the way individual organisms respond to changes in their environment.

The general pattern of development of an organism is laid down within certain channels by the sum total of the genes in the zygote, but the actual end product, the phenotype, is the result of a complex series of reactions between genes and internal and external environmental factors. These reactions can be expressed as an electronic circuit concept whereby genes produce primary products (enzymes) which produce intermediate reactions

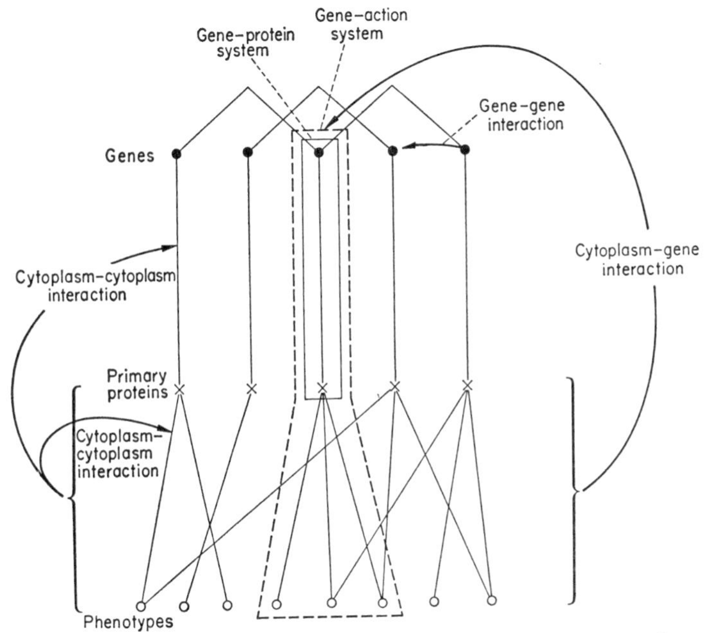

Fig. 5–1 Electronic circuit model of gene action showing interaction and feedback, and indicating that some characters may be the product of the action of several genes (polygeny), while in other cases the same gene may be involved in the production of several characters (pleiotropism). (After WADDINGTON, 1962, *New Patterns in Genetics and Development*, Columbia Univ. Press, New York & London).

and eventually the characters of the phenotype, with a considerable amount of interaction and feedback at all stages (Fig. 5–1). It is important to realize that genes have the *potentiality* to produce a particular character but what is actually produced depends on factors of the genic and external environment during development.

5.2 Genes and characters

If we consider Fig. 5–1 again, we can see that a single gene may be involved in the differentiation of several organs (pleiotropy) and, conversely, several genes may contribute to the production of a single character (polygeny). The effects of a single gene may be small or large. A mutation may cause dramatic changes such as the transformation of a zygomorphic corolla as in *Antirrhinum*, snapdragon, into an actinomorphic one. Another example is found in the genus *Aquilegia* where it has been suggested that the ancestral species possessed flowers without spurs, resembling those found in the Asiatic species *A. ecalcarata*, and the introduction of a spur was probably the result of a single mutation. In crosses between *A. ecalcarata* and species with spurs the ratios found suggest that the presence of a spur in this group of species is determined by a single gene.

If the development of a spur did occur by a single step, it must have dramatically altered the whole evolution of the genus *Aquilegia*. It produced a new type of flower which was immediately isolated from the rest of the population by its necessarily different pollination mechanism. Only pollinating insects with sufficiently long mouth parts were able to extract nectar from the bottom of the long spurs, and the bees which were apparently adapted to these new flowers limited their visits to them. From this ancestral type the whole of the genus *Aquilegia* as we know it today probably evolved.

The *Aquilegia* story illustrates another widespread and important phenomenon—the mutual interdependence of flower and insect. There are many examples in the Angiosperms of the way in which the flower and pollinators have evolved in step with each other, and it is believed that evolution of the flower was in fact largely the result of the development of such a relationship. A good review is given by EHRLICH and RAVEN (1965).

Other examples of the effect of a single-gene mutation include alteration of the pollination system in *Lycopersicon esculentum*, tomato, where the anthers are held in a tube around the stigma by hairs and if they are removed this displaces the anthers from the region of the stigma and cuts down the possibility of selfing. Even the ability of chromosomes to pair may be under simple genetic control.

At the other extreme the effects can be so slight as to merge into the modification of the phenotype by environmental factors. Most of the genes which have contributed to evolution have had small but cumulative effects.

Characters controlled by polygenes (multifactorial inheritance) give rise to continuous variation in populations, randomly distributed so as to give an approximately normal distributional curve (Fig. 5–2a). Environmental variation of the phenotype can also give rise to a normal distributional curve in a population and it is difficult to distinguish between such environmental and genetic variation without experiment.

The effects of *pleiotropic* gene-action can be quite disturbing to the taxonomist. It may produce similar effects on different organs as in the classic example of *Nicotiana*, tobacco. Here the gene causing long petioles is also responsible for related features such as longer calyces with narrow lobes, more acuminate corolla-lobes, longer anthers and more attenuate capsules. Many differences may therefore be due to a single gene and are more likely to be reversed than if they resulted from separate genes. Another effect of pleiotropy is that natural selection may operate indirectly: selection for a character which is due to the pleiotropic effect of a gene will affect other characters which have no direct selective value.

5.3 Phenotypic plasticity

If we return to the effects of the environment on the phenotype we will find that in many organisms the same genotype can produce a range of phenotypes through development in different environments. Such variations are termed *phenotypic modifications* or ecads, and this phenomenon, known as phenotypic plasticity, varies greatly in different species. In some the plants are more liable to be modified by developmental or environmental factors than others which are more stable. The genotype has a range of plasticity which is determined genetically and in some species the range of expression is wide, in others it is narrow.

Not only do species differ in their overall plasticity but so do different parts or organs of the plant. Features such as leaf arrangement and floral structure tend to remain virtually unaffected by different environmental features while others, such as stem height, leaf dimensions and flowering time, can be considerably modified. This is partly due to the different basic patterns of their developmental pathways, but since features with the same pathway in different species vary in their plastic responses it follows that plasticity is a property specific to individual characters in relation to specific environmental factors. Furthermore we may have to consider plasticity in relation to particular environmental factors at different stages of development: in the sunflower (*Helianthus annuus*), for example, there are cultivated and wild forms within the same species. When we grow a large number of plants per given area of the wild form, the number of seeds per plant remains stable but the size of the individual seeds is greatly reduced when compared with plants grown in isolation. Cultivated forms show the opposite response; i.e. in conditions of high density, seed weight remains stable whilst seed number per plant is reduced proportional to the number of plants per given area. This also suggests that we may be able to select for phenotypic plasticity.

Characters which are relatively stable are sometimes called 'good' characters since they can be relied upon for taxonomic comparisions. It is clearly important for the taxonomist to be able to distinguish between the effects of environmental modification and features caused by genotypic differentiation. This is often difficult to establish and the only certain method is by comparative cultivation of the plants concerned. The techniques used for the study of plasticity are (*a*) artificial modification of the environment by means of greenhouses or growth chambers with some degree of control of temperature, light, humidity, (*b*) transplant experiments of two main sorts—cultivation of plants of uniform genetic constitution (clones) in different environments, and cultivation of plants of apparently different genotypes under uniform environmental conditions over a number of years.

Aquatic plants are notoriously plastic although the causes of the wide range of phenotypes often observed are much more complex than previously believed. Plate 1 illustrates the range of phenotypes produced by a single genotype of an aquatic species of *Ranunculus*.

Although a great deal of work has been done on the effects of individual factors such as light, temperature, wind, crowding, soil, etc. (see DAUBENMIRE, 1959, for a good account), they do not operate singly in nature but form part of an intricate, interrelated, environmental complex and it is usually difficult to sort out which factor is responsible for any particular effect.

5·4 Natural selection

Natural selection is a term that covers a large number of different phenomena, all of which are means by which plants become adapted. It operates between different individuals in a population, different local populations, different species, and can be the result of the differential behaviour of many factors such as mortality, fertility, breeding success and so on.

Selection also has to be considered in relation to the type of environment (Fig. 5–2).

5·5 Ecotypic differentiation

We have seen that each species shows its own particular response to different environmental conditions, and it can also be observed that different populations of a species occupying distinctive habitats may show different structural and physiological features, often of an adaptive nature. Experiments by the Swedish botanist TURESSON demonstrated that in several cases these differences persisted when samples of the different

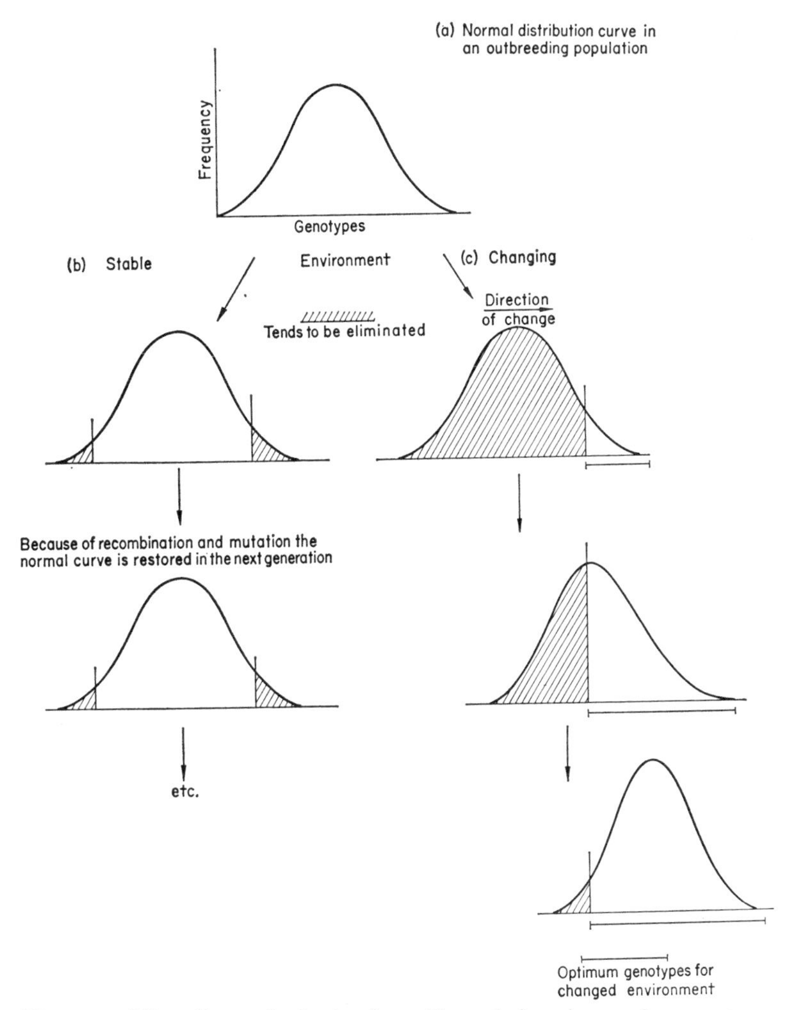

Fig. 5–2 The effects of selection in stable and changing environments.

populations were grown under uniform conditions, indicating that they were due to genotypic differences, not to plastic responses. These genetically different races were often correlated with habitat differences, and moreover similar responses occurred in different species growing under the same or similar environmental conditions. Low-growing or prostrate maritime races (Plate 2) are found in unrelated species which normally show an erect habit. It seems also that when similar environmental factors exist within the area of a species, the same or similar ecological races could be expected to evolve.

TURESSON called these genetically adapted, ecological races, *ecotypes* and similar phenomena have since been described in a large range of species. He also showed that parallel phenotypic modifications were induced by the same environmental conditions that caused the genetically determined ecotypes.

A large-scale series of experiments on genecological differentiation in plant species was undertaken by the American team of CLAUSEN, KECK and HIESEY who established experimental gardens at three stations at altitudes of 100 ft, 4,600 ft and 10,000 ft on a 200-mile transect across Central California which covered a very wide range of ecological and climatic conditions. Various groups including *Potentilla glandulosa* and *Achillea millefolium* were sampled by these and other workers from this transect and it was found that those species which extended across large portions of the transect comprised a series of climatic races or ecotypes, restricted in some instances to ecological zones within the different climates. Some of these races were morphologically differentiated and treated taxonomically as subspecies; in other cases there was continuous variation in morphological features or, as in *Deschampsia caespitosa*, only physiological adaptation.

The large-scale ecotypes of the American school, treated as subspecies taxonomically when morphologically recognizable, contrasted with the small-scale ecotypes of TURESSON which he regarded as taxonomically nearer varieties. Another pattern of ecotypic differentiation was found by GREGOR working in Scotland on *Plantago maritima* which occurs in a wide range of maritime habitats in small, more or less isolated populations. He found that the ecotypic variation pattern was more often continuous than discontinuous and he referred to this continuous ecotypic variation as an *ecocline*. Similar patterns have since been found in other groups ranging from other species of *Plantago* to species of *Eucalyptus*.

A further pattern of ecotypic differentiation has been described by BRADSHAW where the environment forms a patchwork or mosaic of different conditions to which the plants adapt by forming a corresponding mosaic of differentiation. This can happen on a very small scale and ecotypic differences can occur between populations separated by distances as little as 1 metre. This mosaic pattern of ecotypic differentiation has been found in *Agrostis tenuis* in areas of Central Wales where the species is represented by a series of local populations between which isolation is sufficiently effective to permit differentiation to occur in response to the local environmental conditions. No taxonomic recognition of the different variants is possible in such cases. When environmental differences occur over such small distances, genetically different populations may only occur when selection pressures are sufficiently high.

These examples illustrate the key role of natural selection in moulding variability into various units with which the taxonomist has to deal.

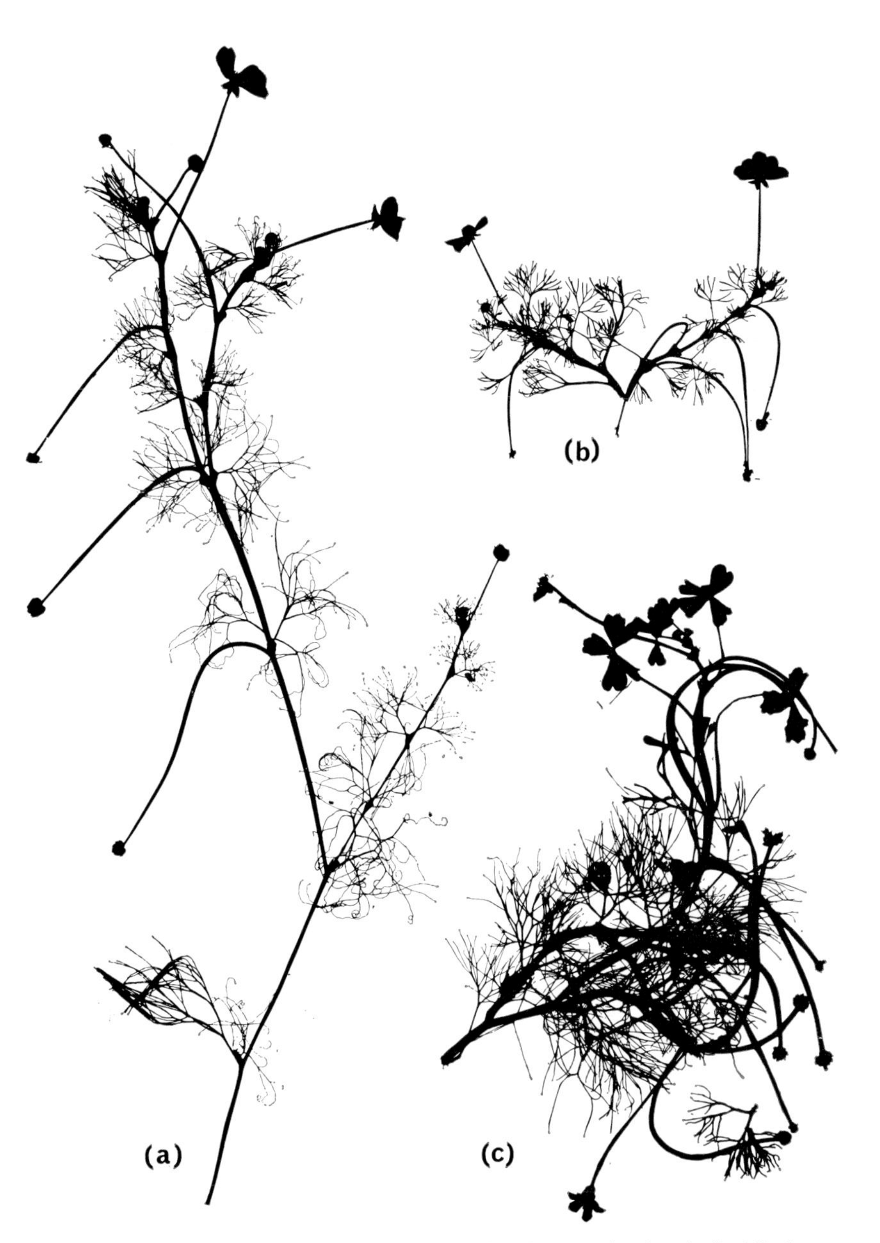

Plate 1. Phenotypic plasticity illustrated by *Ranunculus baudotii*. All the plants shown are ramets of a single clone collected on 20 June 1960 at Zicksee, Austria, and grown at the Munich Botanic Garden. (**a**) Cultivated in 50 cm water; (**b**) cultivated terrestrially; (**c**) cultivated in 10 cm water. (Data supplied by DR. C. D. K. COOK.)

Plate 2. Ecotypes of *Cytisus scoparius*. In addition to the obvious habit differences, the ecotypes show different degrees of hairiness. Time of flowering is also different, as can be seen in the photograph.

Taxonomic Characters 6

Nearly all the activities of systematics involve handling taxonomic characters and their variation. Although it is whole organisms that we classify, it is a selection of their characters which provides the data used in classifications, and how successful our classifications are depends largely on the aspects of the organisms that we select and consider as characters and how we treat them.

6.1 The nature of characters

A character may be generally defined in taxonomy as any attribute referring to form, structure, physiology or behaviour which is considered separately from the whole organism for a particular purpose such as comparison, identification or interpretation. Characters are abstractions and taxonomists deal with their expressions (or states): stem-height is a character, stems 10–30 cm high is an expression of that character.

In practice a character is any feature of an organism or taxonomic group that can be measured, counted or otherwise assessed. Defined in this way biochemical, physiological, cytological and other kinds of data are covered as well as traditional morphological features. The kind of characters employed depends on the class of plants concerned: in Angiosperms mainly morphological features have been used up to the present, while in bacteria mainly chemical test characters are used. The reasons are partly practical, partly traditional. Angiosperms show a wide range of form and structure, and morphological features are therefore easy and quick to appreciate and apply; bacteria on the other hand show much less morphological variation and this is difficult to observe. Thus their biochemistry is employed.

Although an individual organism may possess thousands of potential characters, there is obviously a limit to the number we can use and a selection has to be made for practical reasons. If different taxonomists make different selections of characters from the same plants, the resultant classifications may not be the same. The problem of making a relevant choice of characters is therefore important.

Characters may be simple or complex although this is not always easy to appreciate. An apparently simple feature such as the shape of a leaf can be divided into many separate characters such as length, maximum breadth, breadth at various intervals, length/breadth ratio, base, apex, margin and so on. These components may be under the control of different genes although they may interact. A good practical exercise is to see how many characters can be found to distinguish between the leaves of related species. In different *Rhododendron* leaves, for example, it is not difficult to find at least thirty characters. Again the complexity of a character may depend on the level of magnification at which we observe it: the hairs on a leaf look very different according to whether we use a ×10 lens or a binocular dissecting microscope with magnifications up to ×250.

Characters do not occur uniformly in an organism but in associated suites or groups, such as the different parts of the flower, to take an obvious example. It is not always possible to appreciate these associations of characters in a casual inspection of a plant and often they do not become apparent until a classification has been made. This difficulty affects both the problem of sampling characters and the biological meaning of the character suites. Natural selection operates on character combinations, not on those features of a plant that we choose to regard as characters.

Characters may be scored directly by some form of numerical assessment such as counting, measurement, or angle. This includes qualitative features, such as leaf shape, which can be expressed quantitatively. Other characters such as colour, smell, texture, etc., are not susceptible to direct assessment by numerical means.

Characters may be shown by experience to be 'good' or 'bad' for particular purposes. 'Good' characters are those that are not subject to wide variation in the sample being studied, are not easily modified by environmental factors and are of such a genetic basis that they are unlikely to change readily.

We will now consider some of the kinds of characters that are employed in plant taxonomy.

6.2 Morphology and anatomy

Morphological features have been studied so extensively by botanists in the various classes of plants that it might be assumed there is little left to learn. That this is not so can be easily shown. The fact is that there are so many thousands of species to study that it simply has not been possible to look at them in detail, and many earlier observations have been shown to be wrong or misinterpreted. Much attention is being concentrated today on microscopic or submicroscopic features, the so-called trivial features such as spines, hairs, sculpturing of spores and pollen grains, epidermal structure, etc. which are often used as diagnostic features in the separation of species or genera.

The development of the scanning electron microscope (Stereoscan) which permits the detailed examination and photography of specimens at high magnification and with remarkably great depth of focus has opened up a whole new dimension to the morphologist. It shows up unsuspected detail in features such as fungal spores, pollen grains, leaf-surfaces, etc. An example is given in Plates 3a–c of *Turgenia latifolia*, an umbellifer related to the common carrot. In this group of Umbelliferae characters such as the number of rows and arrangement of the spines on the fruits (mericarps) are used to separate closely related genera, and many taxonomists have felt unhappy relying on such trivial features. In Plate 3(a) we see the complete fruit of *Turgenia* in which the rows of spines on the secondary ridges are clearly visible; examination of the spines under the Stereoscan at magnifications of × 130 (Plate 3(b)) reveals a wealth of detail which is resolved in Plate 3(c) and (d) at higher magnifications with a number of unsuspected elements. In preliminary examinations of the fruits of this and related genera it is already clear that apparently simple features are made up of a series of separate and distinct 'micro-characters' which would appear to reinforce the distinctions made upon the trivial characters alone. Although it is possible, now we know of their existence, to recognize these micro-characters under the light microscope, their interpretation is more difficult and several of them are damaged by the techniques which have to be employed to prepare them for examination.

Details of the epidermis in the grasses have been studied intensively by light microscopy in recent years and it has been found that the major subdivisions of the family Gramineae (Festucoideae, Panicoideae, etc.) are characterized by the shape of the siliceous cells present in the epidermis and the presence or absence of bicellular hairs and their form when they occur. In addition some features are restricted to smaller groups such as tribes or genera; and some elements are common to all groups. Thus for each species a dermogram may be constructed giving the distribution and shape of each epidermal element.

Stomata and associated epidermal cells also provide taxonomically useful characters in many groups. A recent survey of the stomata and their subsidiary cells in the leaves of the monocotyledons has shown that each genus tends to be constant in these features and their pattern of development is characteristic for each of the cell complexes. A useful review of epidermal features is given by STACE (1965).

Anatomical features are widely used in systematics—for identification, for placing anomalous groups in a satisfactory position in classifications, and for indicating patterns of relationship that may have been obscured by superficial convergence in morphological features. Floral anatomy received a great impetus by the development of rapid clearing techniques for the study of floral vascular systems which complemented the classical methods of staining serial sections. Using such clearing and bleaching methods MELVILLE (1962) has surveyed a range of Angiosperm families and proposed a new theory of the nature and origin of the Angiosperm flower—the *gonophyll* theory which holds that the basic structure of the ovary was a leaf bearing a dichotomous fertile branch on its midrib or petiole. The whole question of the evolutionary origin and nature of the Angiosperm flower has been reinvestigated by several botanists recently and the results are reviewed by MEEUSE (1965).

Wood anatomy has been used successfully in many groups. It has helped to establish the systematic position of the 'primitive' vesselless Angiosperm families Winteraceae, Trochodendronaceae and others, and contributed to the evidence for treating the genus *Paeonia* as a separate family from the Ranunculaceae in which it is often included. Perhaps the most spectacular contribution has been the remarkable series of studies by BAILEY and his associates which established a number of trends of evolutionary specialization in stelar structure (BAILEY, 1954). These trends were in size, structure and pitting of the tracheary cells and together they constitute one of the best documented, most complete and reliable evolutionary sequences known in the plant kingdom. What is more, there is little doubt about the direction in which the trends should be read and they are apparently irreversible.

Such studies have also helped to establish that the *Magnolia* type of plant (Magnoliaceae, Winteraceae, etc.) with large showy flowers comprising many, spirally-arranged parts, are less specialized than the Amentiferae—those families such as the oaks, hazels, willows, etc. with much reduced, often unisexual, flowers arranged in catkins. The wood anatomy of the *Magnolia* group is generally less specialized than those of other Angiosperm families, while that of the Amentiferae is often quite specialized and varies considerably between the different families. A valuable summary of the role of anatomy in taxonomy is given by CARLQUIST (1961).

6.3 Pollen, spores and embryos

Features of pollen and spores are being increasingly used in systematic work. In addition to size and general shape, the principal characters of pollen grains are number and position of the furrows, the number and position of the apertures and the details of sculpturing of the wall (sporoderm), especially the outer layer (exine). The details of the exine are such that it can be used in plant identification much in the way that fingerprints are used for identification of criminals. Grains and spores are highly resistant and are found preserved in fossils in peat and lake sediments thus providing evidence about vegetational history back to Cambrian times.

Pollen grains may be observed by ordinary microscopy, phase-contrast, or in some cases polarizing or fluorescence microscopy. Fine detail, especially of the intine and cytoplasm, requires the use of the ultra-violet microscope, but when it comes to the finest details such as surface relief of the grains and spores, electron microscopy is necessary. It is in the study of the ultrafine structure and stratification of the sporoderm by means of ultrathin sections examined by electron microscopy that the most exciting advances have been made in recent years. There are still many problems of interpretation involved in considering the details of the ultrastructure of grains and spores, but already data of considerable taxonomic value have been obtained in the Compositae (STIX, 1960). The fact that fossil pollen retains most of its structural detail will be of great significance when evolutionary trends in the development of the sporoderm are firmly established. An excellent review of the systematic applications of palynology is given by ERDTMAN (1963).

Comparative embryology has made a smaller contribution to systematics than palynology or anatomy. Data are scarce, techniques are often difficult and interpretation may be difficult. Outstanding examples of embryological data applied to taxonomy are CAVE's studies on generic delimitation in the Liliaceae and REEDER's work on the main types of embryo found in the Gramineae. A summary of advances in this difficult field is given by MAHESHWARI (1963).

Biochemical Systematics 7

Chemical characteristics of plants have been noted and used by taxonomists for centuries. Examples are the characteristic essential oils in labiate genera such as thyme, catmint, mints, the turpentines of pines and the oils and other constituents of umbellifers. The two major divisions of the Compositae, the Tubiflorae and Liguliflorae are distinguished by what is essentially a chemical character—the presence or absence of latex. But it is only in the last ten years or so that biochemical systematics has come to occupy a major role. The two main reasons for this have been the development of rapid and efficient screening techniques such as chromatography and electrophoresis, and, as a result of the rapid identification of large numbers of organic compounds by these methods, the realization that they have a wide systematic value and can contribute to the solution of many taxonomic problems.

The dominant approach today is the systematic surveying of plant groups for secondary compounds of low molecular weight which are the by-products of the major metabolic pathways. These substances such as alkaloids, non-protein free amino acids, flavonoids, glycosides and terpenoids have an irregular distribution in the plant kingdom and their function is unknown in the majority of cases, although they often have a remarkable physiological effect on higher organisms and have attracted the attention of chemists and pharmacologists for many years. They may occupy some kind of protective role in plants or act as buffers for other important metabolic processes. The fact is, however, that we do not yet know much about their biological significance to the plant and they appear to be expressions of metabolic virtuosity and provide the systematist with a rich source of data for comparative and interpretative purposes.

Some low-molecular-weight substances are of such wide occurrence in plants that their taxonomic value is negligible; e.g. glucose, the twenty protein amino acids. Others such as xanthine are ubiquitous products of

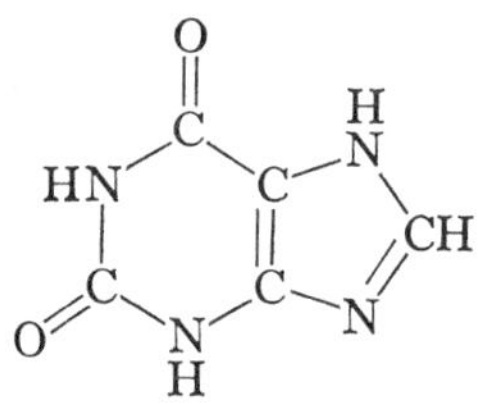

Xanthine, ubiquitous in plants.

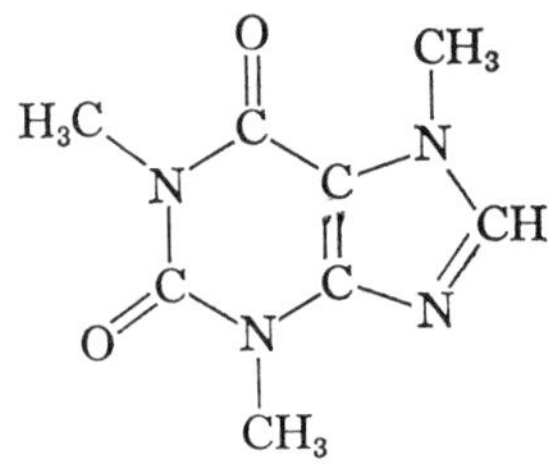

Caffeine, the active principle of coffee, tea, cocoa, etc.

metabolism, while the purine caffeine which has a similar structure that may have arisen from it by a single mutational step is found in a number of apparently unrelated groups of Angiosperms and presumably arose independently a number of times through parallel chemical evolution.

Other substances are extremely restricted in distribution such as the isoflavone iridin known only in the *Pogoniris* section of *Iris* and the free amino acid, lathyrine, known so far only in the genus *Lathyrus*. In both *Lathyrus* and *Vicia* it has been shown that different groups of species are characterized by the overall distribution pattern of amino acids as well as by their morphological characteristics. Another interesting distribution pattern is found in the recently discovered class of compounds, the biflavonyls which are restricted to the Gymnosperms (but not in the Pinaceae) and to the genera *Psilotum*, *Selaginella*, *Casuarina* and *Viburnum*!

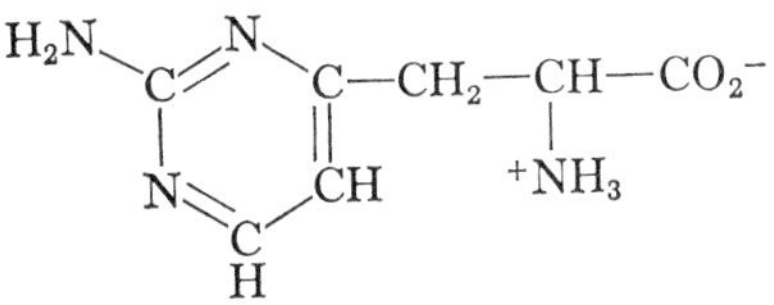

Lathyrine, an amino acid found only in *Lathyrus*.

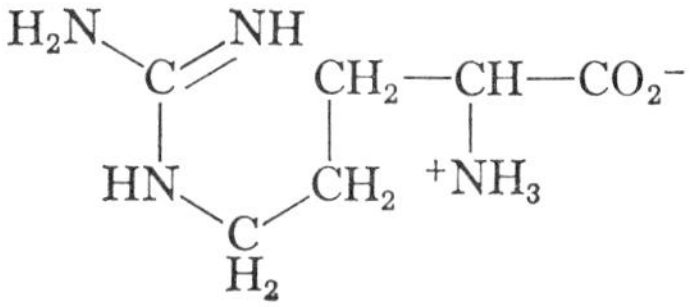

Arginine, one of the twenty protein amino acids.

Perhaps the most striking example concerns the taxonomic distribution of the red and yellow pigments, the betacyanins and betaxanthins. These are confined to ten Angiosperm families which have been treated collectively as the Centrospermae; they do not occur together in any plants with the anthocyanins which are the normal pigments found in most other Angiosperm families (although other classes of flavonoids do occur in them), and they are unrelated both chemically and biosynthetically to the normal pigments.

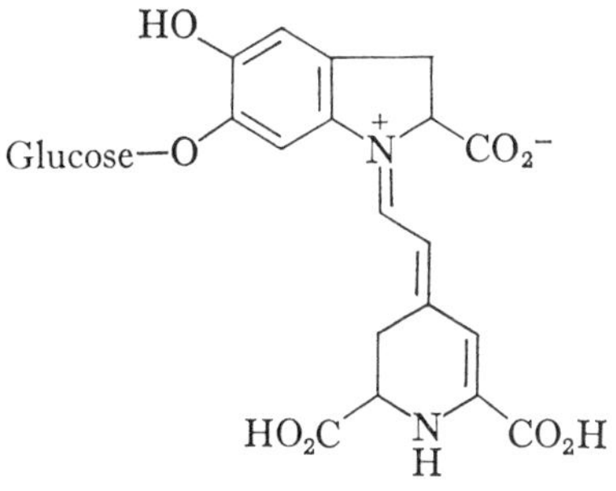

Betanin, the beetroot pigment; a betacyanin.

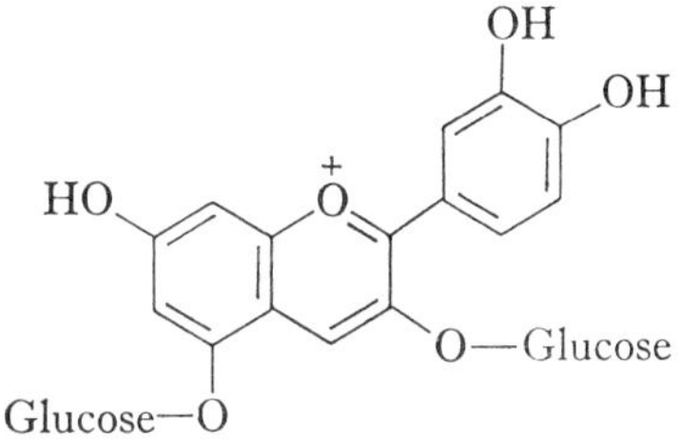

Cyanin, the cornflower pigment; an anthocyanin.

The taxonomic significance of the betacyanins and betaxanthins depends not just on their unique chemical structures nor on their restriction to the Centrospermae but also on the fact that they are mutually exclusive with the widely distributed anthocyanin pigments. If we take into account as well morphological, anatomical and embryological evidence there are good grounds for limiting the Centrospermae to those families which contain betacyanins (Chenopodiaceae, Portulacaceae, Aizoaceae, Cactaceae, Nyctaginaceae, Phytolaccaceae, Stegnospermaceae, Basellaceae, Amaranthaceae and Didieraceae), while families containing anthocyanins and previously included in the Centrospermae should be placed in a separate order, the Caryophyllales (Caryophyllaceae, Illecebraceae, Molluginaceae).

The chemical evidence suggests that the betacyanins and betaxanthins had a single evolutionary origin, possibly even before the general appearance of anthocyanins in the angiosperms and that the betacyanin-containing families therefore constitute a distinct phyletic lineage. If this is so, then it counters HUTCHINSON'S arrangement of these families into widely separated orders according to their woody or herbaceous nature. HUTCHINSON'S treatment would require the hypothesis that the betacyanins evolved independently in these different orders, which is highly implausible in the face of the biochemical evidence.

7.1 Chromatography and hybrids

The patterns of flavonoid compounds extracted from various parts of a plant may allow us to obtain a typical chromatographic profile for an individual, population or species (Fig. 7–1). This technique has proved

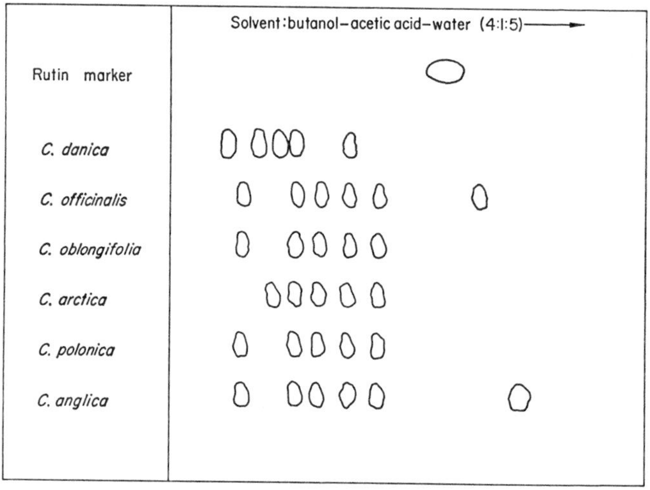

Fig. 7–1 Chromatogram of flavones in petal extracts of *Cochlearia* species.

extremely valuable in the analysis of hybridization situations, especially where the morphological features of the species and hybrids are difficult to interpret or where there is a large degree of backcrossing. An outstanding example is found in the genus *Baptisia* where each species has a distinct chromatographic pattern of flavonoid compounds, and chromatograms of the hybrids show the compounds of both parental species (Plate 4).

This additive expression of flavonoid patterns in hybrids has been recorded in several other groups. Another remarkable case has recently been described in the fern genus *Asplenium*. There the allopolyploid hybrid *A.* × *kentuckiense* (which it is thought combines the genomes of three diploid species, *A. rhizophyllum*, *A. montanum* and *A. platyneuron*) was shown when analysed chromatographically to present a profile in which the patterns of each of the three species which are believed to have been involved in the origin of this polyploid hybrid (Fig. 7–2) are added together. This chemical

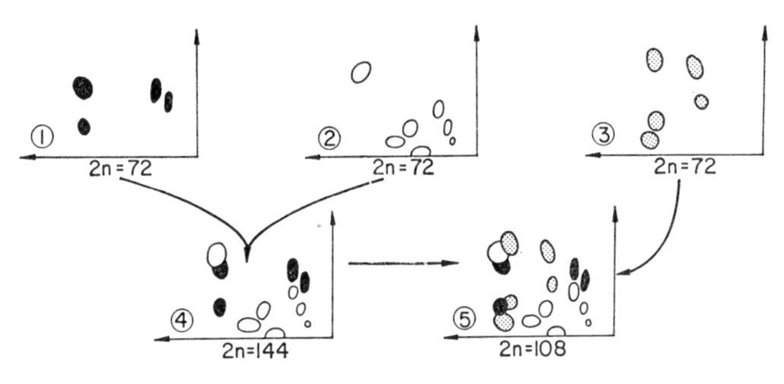

Fig. 7–2 Diagrammatic representations of two-dimensional chromatograms of three species of *Asplenium* and their hybrids, showing the complementation of flavonoids referred to in the text. (1) *A. rhizophyllum*; (2) *A. montanum*; (3) *A. platyneuron*; (4) bigenomic allopolyploid (*A. rhizophyllum* × *A. montanum*); (5) trigenomic allopolyploid, *A.* × *kentuckiense* (*A. rhizophyllum* × *A. montanum* × *A. platyneuron*). (After SMITH and LEVIN, 1963 *Amer. J. Bot.*, **50**, 952–958.)

study agreed with views concerning the relationship of *A.* × *kentuckiense* reached by other workers using morphological and cytogenetic data.

Chromatographic analysis of hybrids is not always successful, especially if the parental types do not possess distinct chromatographic profiles. On the other hand it may be more successful than morphological analysis in cases where the hybrids closely resemble one or other of the parental species.

The techniques of paper chromatography, although they require skill, are relatively simple and rapid, and they only need a small amount of leaf material. Gas chromatography, where the extracts are volatilized and run as a gas through a liquid column instead of on paper, requires more expensive apparatus but can be largely automated, the results being presented in the form of a graph in which the peaks represent the abundance of molecules with different features. It is employed for the screening of oils as in mints, or terpenes as in the pines. An almost complete survey of the turpentine composition of the 100 or so species of the genus *Pinus* has been undertaken by MIROV and his associates. The chemical composition of oils has been extensively studied in the genus *Eucalyptus* and the results have been of considerable taxonomic value.

It is not even necessary to use fresh material for chromatography. We have recently surveyed with excellent results, a wide range of species of Umbelliferae for phenolic compounds using herbarium material up to 100 years old.

7.2 Macromolecular systematics

Until recently the main source of systematic information about proteins was derived from serology. Most of the serological work on plants has employed the precipitin reaction in which protein extracts from a given plant are injected into a rabbit, causing the formation of antibodies in its blood serum. The serum containing antibodies (antiserum) is then mixed with a suspension of the protein to be tested (the antigen) and the antibody and antigen react forming a precipitate, hence the term precipitin reaction. The amount of precipitate can be measured and it is found that the reaction is specific for the antigen used to induce the formation of the original antibodies (homologous reaction), and this is regarded as a *standard* or *reference reaction* against which reactions with antigens from other plants can be compared (heterologous reaction). The strength of the reaction is regarded as a measure of the protein similarity of the samples and therefore to some extent of the plants being compared. There is generally a close relationship between the results obtained from the precipitin test and general ideas on affinity as judged by morphological and other data. Much recent work in this field has been undertaken at Rutgers, New Jersey, by FAIRBROTHERS and JOHNSON, and their estimates of serological correspondence have been valuable in assessing relationships in such groups as the Magnoliaceae, Cornaceae and Gramineae (see LEONE, 1964).

A more recent development in serology is the use of gel diffusion and electrophoretic methods in which the antigens and antisera diffuse electrophoretically towards one another in an agar gel, forming bands of precipitate. This method allows qualitative estimates of the structure of the proteins to be made. A study of Mexican potato species has been made at Birmingham using this technique with some success (GELL *et al.* 1959). Hybrids of *Phaseolus vulgaris* × *P. coccineus* have been identified by immuno-electrophoresis and it was found that the F_1 plants showed complementation of the precipitate bands of the parents, analogous to the additive chromatographic profiles mentioned in *Baptisia* hybrids above.

The techniques of serology are difficult and there are many acute problems involved in interpreting the results. Moreover only limited areas of the surfaces of the antigen molecules are measured for biological similarity and the reactions of only a few genes are involved.

Although very little information has yet been obtained about the detailed structure of macromolecular compounds such as proteins and deoxyribonucleic acids, techniques similar to those used for the analysis of secondary compounds, such as gel electrophoresis, have been employed recently for the detection of protein patterns or bands. By some of these methods rapid 'finger-printing' of proteins is possible. There have been few applications to plant systematics so far, but work undertaken by BOULTER and his associates at Liverpool on protein bands obtained from the albumin fraction of seeds of various genera of the Leguminosae has shown that the techniques may be of great value in supporting taxonomic arrangements or suggesting where revision ought to be considered.

Knowledge of the detailed protein structure of different organisms is expected by some biologists to lead to a protein taxonomy in which the complete amino-acid sequences for all the proteins produced by each species would form the basis for classification since it would amount to a complete genetic description of the species. Automatic determination of amino-acid sequences and comparison of them by electronic computers is something for the future. To date the complete amino-acid sequence has been worked out for only one plant protein (cytochrome *c*) and tentatively for another (papain). In time it will be possible to work out trends of molecular evolution and protein 'phylogenies'; it has even been suggested that from the basis of existing polypeptide chains it will eventually be feasible to reconstruct some of the polypeptide chains of ancestral organisms and make some informed deductions about their characteristics. This field, known as chemical palaeogenetics, is reviewed by ZUCKERKANDL (1965).

7.3 DNA-hybridization

The ability of nucleic acids to form 'molecular hybrids' by base-pairing opens up an enormous, new and exciting field which may allow us to make direct estimates of overall genetic similarity (HOYER et al. 1964). Strands of DNA extracted from an organism are untwined and then allowed to hybridize *in vitro* with DNA or RNA extracted from another organism, and the homologies between them can be expressed quantitatively by measuring their absorption in ultraviolet light. This DNA-hybridization technique has been employed almost entirely in animals and micro-organisms so far, and only preliminary results of the first experiments with plants have been published by BOLTON and his associates at the Carnegie Institution of Washington. These results are somewhat disconcerting; for example it was found that only half the nucleotide sequences in the DNA of the vetch *Vicia villosa* are homologous with those in the pea, *Pisum*, while only a fifth are common to beans (*Phaseolus vulgaris*) and peas. But it is clear that if the claims made for the technique are substantiated, it will provide the only method for obtaining a direct and objective assessment of the degree of relationship between higher taxonomic groups.

The following books review recent developments in the field of biochemical systematics—ALSTON and TURNER (1963), LEONE (1964), SWAIN (1963, 1966). HEGNAUER'S vast systematic survey of chemotaxonomic data, *Chemotaxonomie der Pflanzen* (1962–6) should also be mentioned.

Weighting: is A more important than B? 8

In the previous chapters we have considered some of the classes of data that are employed in systematics. Obviously they are not all of equal value for purposes of comparison, identification, classification, interpretation and so on: taxonomists therefore weight characters. To weight a character means giving it greater or lesser importance than other characters. There are several different ways of weighting and the skill of a taxonomist depends to a large extent on his use of weighting.

Since there is virtually an unlimited number of characters available, in theory, a selection of a smaller number of them has to be made by the taxonomist when constructing a classification, according to criteria such as ease of observation, availability, experience, etc. We may call this *selection* weighting. At the same time other characters are rejected because they are too plastic in response to environmental fluctuations or are considered unreliable for other reasons. Since this has the effect of giving greater importance to those characters that are chosen, this is called *rejection* or *residual* weighting.

Apart from these necessary procedures of selection and rejection there are two main forms of weighting—*a priori* and *a posteriori*. The former means that some characters are selected because they are regarded as of particular importance—because they are genetically stable or conservative, because they are known to be good diagnostic features in other groups, because they are known or assumed to indicate phylogenetic relationships. Taxonomists have used this form of weighting for centuries: insistence on the use of floral characters originated from a belief in their physiological importance; HUTCHINSON'S division of the Dicotyledons into predominantly woody and herbaceous lines is an extreme form of *a-priori* weighting based on the belief that habit is an indication of major evolutionary divergence. Those taxonomists who believe that morphological features are more important in classification, rather than more practical, are employing *a-priori* weighting, and so is nearly every taxonomist when he decides that a particular character or set of characters in a group he is working on is going to be particularly valuable in making a classification *before* he has any evidence to justify his belief.

Every new source of data applied to taxonomy tends to be given *a-priori* weighting, largely because of its very novelty. Thus in the early days of cytotaxonomy there was a tendency to regard chromosome number as a much more important feature of a plant than morphological characters, and so plants possessing different chromosome numbers were sometimes placed in separate groups although their morphological and other characters did not agree with this separation. Often the procedure was successful but for reasons other than those believed. Similarly biochemical characters are often described as more basic or privileged than other kinds of features and tend to be given *a-priori* weighting. Certainly they are particularly valuable since they are often stable, unambiguous and less susceptible to environmental change than other features; chromosome numbers may have similar advantages. The very fact that we have subdisciplines known as chemotaxonomy, cytotaxonomy, etc. reflects the tendency to weight these classes of data, although the terms refer to the specialization of techniques and approach.

As we have seen, the advantages possessed by cytological, biochemical and other features provide us with a reason for *selecting* them for use but not for giving them any further weighting beforehand. *After* a classification has been made, it is often found that they show a high correlation with other features of the groups in the classification and can be considered good 'marker' or diagnostic characters. In many cases this is found not to be true, and since we cannot predict the results beforehand there appears to be no justification for *a-priori* weighting of any kind of data. There are, it seems, no short cuts to classification and we cannot proceed on a hit-or-miss basis by choosing characters which we hope and believe will be important.

After groups are constructed it is found that some characters show high correlations with others and these characters are given *a-posteriori* or *correlation* weighting. Since we cannot know which these characters are until after the groups have been made, correlation weighting is a way of being wise after the event. The taxonomist makes these character correlations, either by the integrating capacity of his own brain (so-called intuitive or neural classification) or by tabulation and mathematical procedures or by using an electronic computer. Although the integrating capacity of the human brain is vastly superior to any form of computer yet devised, we are far from clear what the mental processes involved are, that is, which characters are compared with which, in groups or singly. When using morphological features only, the taxonomist is able to make these correlations fairly efficiently as the great array of taxonomic classification used today clearly shows. If however he has to integrate both morphological and non-visible features such as chemical characters, the method breaks down unless it is divided into stages—correlating the morphological features first and then seeing how the other characters fit in. This obviously introduces a great deal of bias and the procedures of numerical taxonomy (Chapter 9) have been designed to overcome these problems.

The experienced taxonomist using his perception and experience is often able to judge what characters are likely to be useful in the construction of a group. He uses his general knowledge of the distribution of characters in other groups, assesses the known or assumed variation of the various kinds of character available and, by choosing a few, makes a tentative delimitation of groups. If the resultant groups show good character correlations the procedure has been successful and the characters initially chosen are given *a-posteriori* weighting. This method is therefore a form of model testing and it involves treating all the characters *initially* selected as being of equal value until shown otherwise. Equal-weighting (or non-weighting) of characters is considered in Chapter 9.

Computers and Taxonomy 9

The application of high-speed electronic computers to taxonomy represents one of the major developments during the last few years and it is likely to have increasing repercussions on many of the procedures and concepts of various fields of systematics.

Computers are being increasingly employed in the development of quantitative methods of classification—the field known as *numerical taxonomy* or *taxometrics*. Electronic data processing (EDP) is also being applied to procedures in systematics concerned with the recording, storage and retrieval of information which occupies the greater part of the taxonomist's time.

9.1 Numerical taxonomy

This is defined as the numerical evaluation of the similarity between groups of organisms and the ordering of these groups into higher-ranking taxa on the basis of these similarities. The role of the computer is to work out the quantitative comparisons between organisms in respect of a large number of characters simultaneously; in other words it does the arithmetic for us, but at high speed and without bias. With a computer it is now possible to make calculations of similarity between organisms which would have been beyond the capacity of a taxonomist previously.

Numerical taxonomy is based on phenetic evidence, that is to say on the similarities shown by observed and recorded characters of taxa, not on phylogenetic probabilities. Not only is it empirical but it is operational. This means that it is based on statements and hypotheses so formulated that they can be tested by observation and experiment. Thus when we talk about relationship between taxa we have to say by what logical criteria relationship is measured so that they can be communicated to other people and put into practice by them. Since numerical taxonomy is operational in this sense, it is divided into a series of repeatable steps, allowing its results to be checked back step by step.

The logical steps involved in numerical taxonomy are as follows:

1. CHOICE OF UNITS TO BE STUDIED. The first task is to decide what kinds of units to study—these may be individuals, lines or strains, species, etc. It is important that the units should be as representative as possible of the organisms being considered. The entities of lowest rank in any particular study used are called Operational Taxonomic Units (OTU's).

2. CHARACTER SELECTION. As wide a selection as possible is made of the characters of the OTU's; at least 50 and preferably 100 or more are needed to produce a fairly stable and reliable classification, although research is

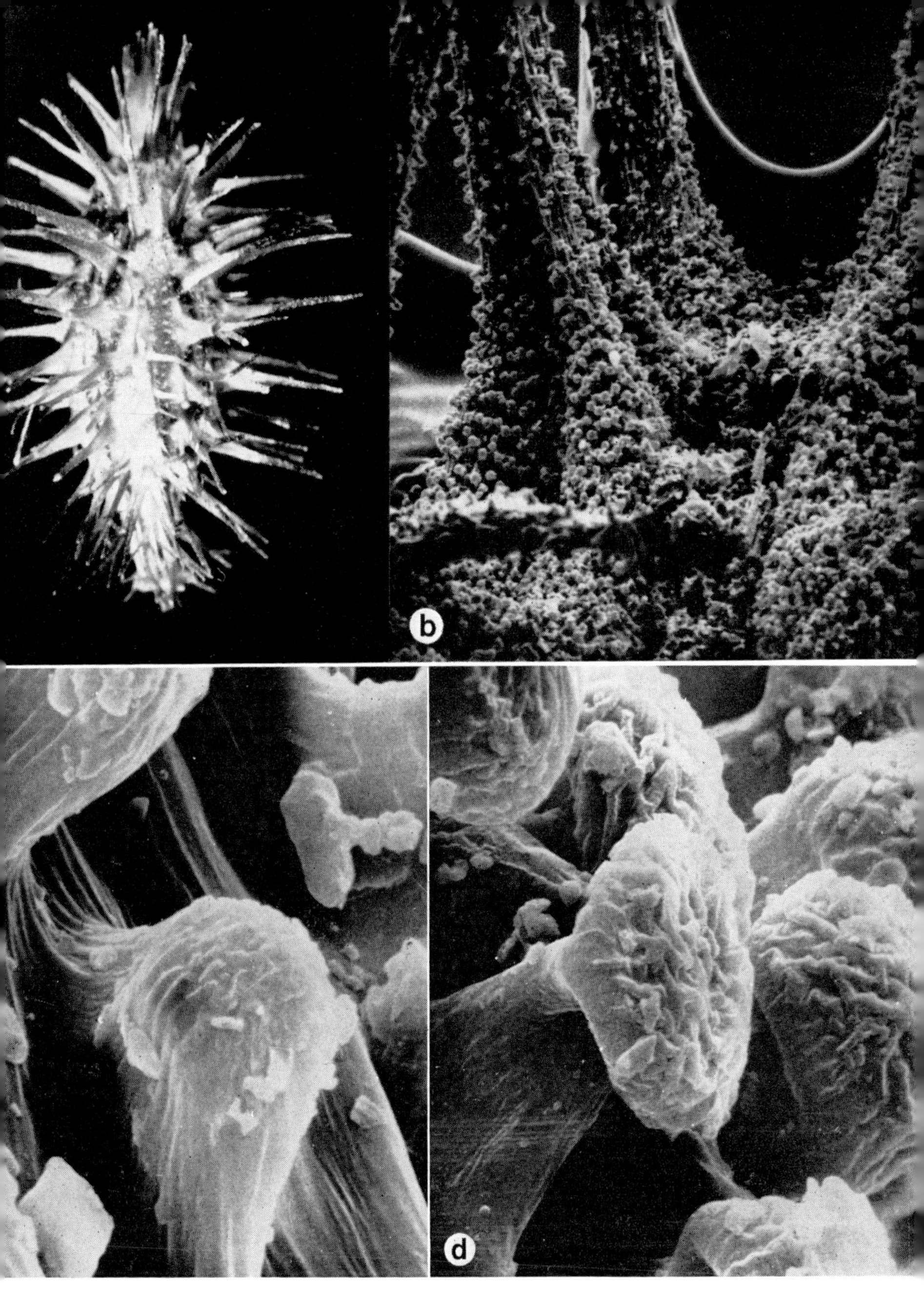

Plate 3. (**a**) Fruit of *Turgenia latifolia* (Umbelliferae). × 10. (**b**) Scanning electron micrograph of the base of the spines of *T. latifolia* shown in (**a**), × 250. (**c** and **d**) Scanning electron micrograph of two types of surface ornamentations on the spines of *T. latifolia* shown in (**b**), × 3,650. (**b, c** and **d** by courtesy of the CAMBRIDGE INSTRUMENT CO.)

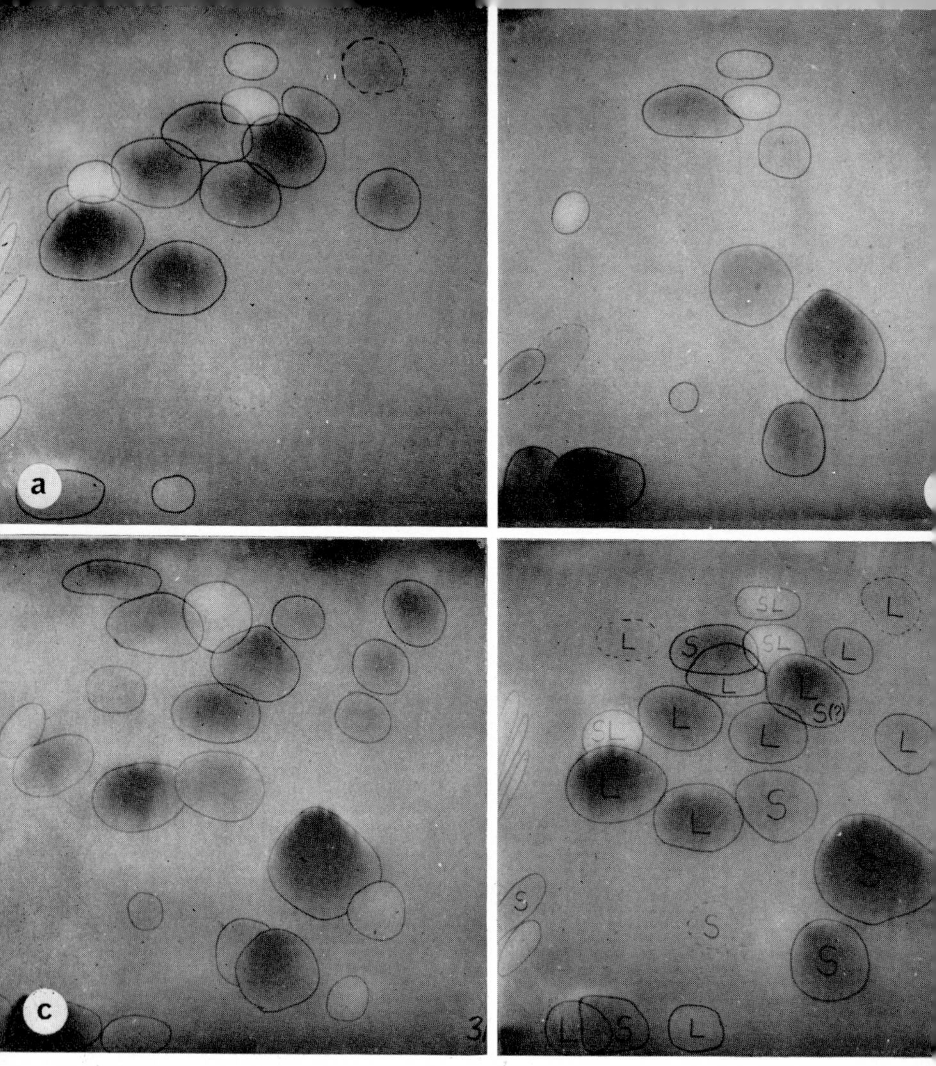

Plate 4. Two-dimensional chromatographic patterns of leaf extracts from *Baptisia*. (**a**) *B. leucantha*; (**b**) *B. sphaerocarpa*; (**c**) pattern of *in vitro* F_1 hybrid *B. leucantha* × *B. sphaerocarpa* (i.e. the pattern obtained by mixing in the test tube, prior to chromatography, the extracts of both *B. leucantha* and *B. sphaerocarpa*); (**d**) pattern of natural F_1 hybrid *B. leucantha* × *B. sphaerocarpa*. The letters *L* (for *leucantha*) and *S* (for *sphaerocarpa*) refer to the species-specific compounds found in the two species. (From ALSTON and HEMPEL, 1964, *J. Heredity*, **55**, 267.)

still going on into the question of how many characters should be used and whether there is an asymptote of information as further characters are added.

The characters used in numerical taxonomy have to be broken down into unit characters. A *unit character* is a character of two or more states which cannot logically be further subdivided. We have already seen in Chapter 6 how difficult it can be to decide how a character can be divided up and defined since this depends largely on our skill and perception.

The characters selected are then coded. All-or-none is the simplest form of coding whereby a character is divided into two states (+ or −). When no data is available the symbol O or NC (no comparison) is used. Multistate coding is when characters are divided into several states, 1, 2, 3, 4, 5, etc., each state representing an equal division of a continuous variable.

All the characters selected are treated as of equal value and not given special *a-priori* weighting. There is no reason why weighting should not be introduced at this stage if there is logical justification for it, but most numerical taxonomy employs equal weighting and is often called Adansonian after the eighteenth-century botanist MICHEL ADANSON who followed similar principles.

The information is then presented in a $t \times n$ table or data matrix consisting of t OTU's scored for n characters (Fig. 9–1).

Characters	Taxa (OTU's)			
	A	B	C	D
1	+	+	−	NC
2	+	+	+	+
3	+	+	+	−
4	−	+	NC	NC
5	+	+	+	+
6	+	+	−	+
7	+	+	−	NC
8	NC	−	+	+
9	+	+	+	+
10	+	+	+	−
11	+	NC	−	NC
12	+	+	+ −	

Fig. 9–1 Coded data table ($t \times n$ table). See text for further explanation. (After SNEATH, 1962, in *Microbial Classification*, edited by AINSWORTH and SNEATH. University Press, Cambridge.)

3. MEASUREMENT OF SIMILARITY. Overall similarity (S) is calculated by comparing each OTU with every other and is usually expressed as a percentage, 100 per cent S for identity and O per cent S for no resemblance. A similarity table or matrix is then constructed tabulating the S coefficients, one for each OTU (Fig. 9–2(a)). Other techniques may also be employed.

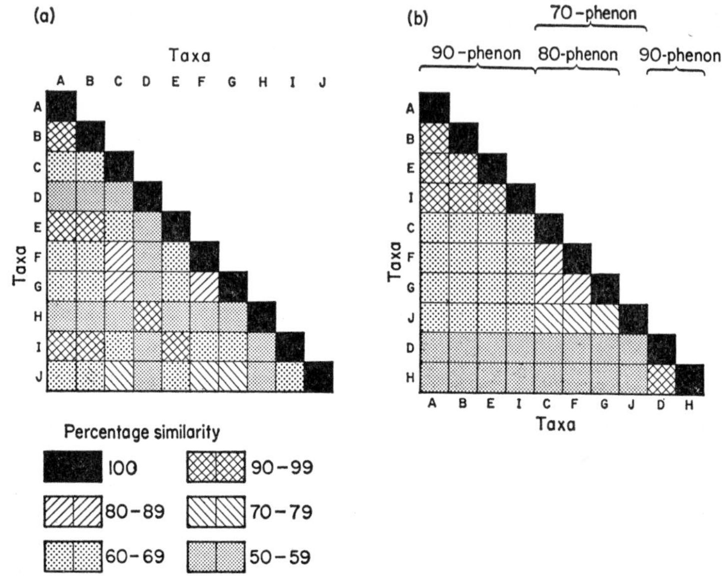

Fig. 9–2 (**a**) Schematic diagram showing a matrix of hypothetical similarity coefficient between pairs of groups (taxa); the magnitude of the coefficients is shown by the depth of shading. (**b**) The same coefficients arranged by placing similar taxa next to each other; this gives a triangle of high similarity values. Phenons are groups of desired rank. (After SNEATH, 1962, in *Microbial Classification*, edited by AINSWORTH and SNEATH. University Press, Cambridge.)

4. CLUSTER ANALYSIS. The similarity matrix is now rearranged so as to bring together into clusters those OTU's whose members have the highest mutual similarity. This can be done by various methods and allows related taxa or groups to be recognized (Fig. 9–2(b)). These clusters are called *phenons* and can be arranged hierarchically in a tree-diagram or dendrogram (Fig. 9–3). By arbitrarily choosing fixed levels of similarity as appropriate to particular ranks, we have an objective means for applying rank uniformly throughout a group which is classified by numerical means.

5. DISCRIMINATION. Having made our classification we can now go back and re-examine the characters to find out which are most constant and therefore most valuable for keying and diagnosis. In other words we have let the computer show us which characters have *a-posteriori* weighting.

The above is a greatly oversimplified account of the procedures of numerical taxonomy; detailed discussion is given by SOKAL and SNEATH (1963). Modifications and refinements are constantly being published and the whole field is in active development. Most of the published applications have been in micro-organisms such as bacteria, but several numerical classifications of groups of higher plants have been produced.

There is much criticism of numerical taxonomy by those who feel that classifications produced by a computer are limited in value since they rely upon a machine to make automatic calculations instead of the sensitive

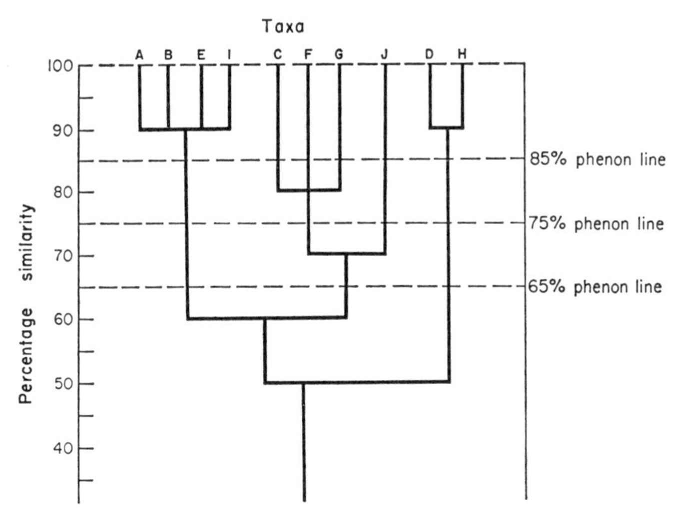

Fig. 9–3 A dendrogram representing the hypothetical hierarchy of groups (taxa) obtained from Fig. 9–2(**b**). The ordinate indicates magnitude of similarity coefficient at which stems join to form higher ranking groups. Horizontal lines delimit groups of equal rank (per cent phenon lines). (After SNEATH, 1962, in *Microbial Classification*, edited by AINSWORTH and SNEATH, University Press, Cambridge.)

judgements of the experienced taxonomist. This seems scarcely justified since the computer performs the time-consuming arithmetic after the taxonomist using all his skill and perception has decided what characters are to be computed. Another major criticism is directed at the equal weighting of the characters employed but, as we saw earlier, the computer in fact works out for us which characters should be weighted. If we weight characters before they are fed into the computer, this implies that we know beforehand what the classification will be like.

Apart from the production of classifications, numerical techniques will allow us to evaluate phylogenetic trends and evolutionary rates (which are assessed by measuring phenetic changes) in a much more precise manner than has previously been possible, by treating them quantitatively. Computational methods are being developed for reconstructing probable branching sequences in phylogeny. It is only with computers that alternative hypotheses about the most probable course of evolution in particular groups can be made in the absence of historical data (see HEYWOOD and MCNEILL, 1964).

It is without question that computers will be needed to handle the newer classes of data being provided for taxonomists by molecular biologists, biochemists and others if they are to take their place alongside the traditional data from morphology. It is the absence of abundant morphological detail which can be quickly assessed and integrated by eye that is largely responsible for the rapid development of numerical classification of bacteria (SNEATH, 1964).

A final point arising from numerical classifications is that they need not necessarily be of the hierarchical, box-within-box type. Although we are used by tradition to the hierarchical form of classification, some taxonomists believe that the unbiased representation of the similarities between organisms is more important and would allow us to have overlapping taxa, that is where a single taxon could belong to more than one higher taxon. The taxonomic hierarchy is largely a mnemonic device and if all information about organisms in relation to all others can be made immediately available by electronic data processing techniques, the need for a hierarchy and even names would largely disappear. Already non-hierarchial classifications are being produced, such as the so-called non-Linnaean taxonomies of DUPRAW (1965).

9.2 Documentation and data processing

Much of systematics is concerned with recording, cataloguing, filing, storing and retrieving information about plants in the form of handbooks, Floras, labels, card indexes, maps, etc. The present apparatus of taxonomy, comprising the binomial system, Latin nomenclature and the hierarchical arrangement of categories provides one of the best information retrieval systems yet devised by man. Yet the sheer bulk of the rapidly accumulating additional information of all sorts which is relevant to taxonomy is such that electronic data-processing equipment will be required to handle it so that it can be made readily available for classifications and other systematic applications. One pioneering step in this field is the *International Plant Index* (GOULD, 1962) which is an attempt to produce a catalogue of plant genera and species, employing an 80-column punched card and which it is aimed will eventually provide a unique reference number for each plant species. The recently published *Atlas of the British Flora* (PERRING and WALTERS, 1962) is an excellent example of the preparation of distribution maps by machine methods, an important feature being that the data employed in the production of each map are stored on punched cards and can be retrieved on request. A review of the whole field of electronic data processing in taxonomy is given by SOKAL and SNEATH (1966).

Chromosomes, Taxonomy and Evolution 10

One of the most important influences on taxonomy during the past few decades has been the contribution of cytology, either alone (*cytotaxonomy*) or interwoven with genetics (*cytogenetics*).

Cytological features such as chromosome numbers and morphology can be used like any other kind of comparative data; chromosome behaviour and structure at meiosis contribute to our understanding of the evolution and relationships of populations.

The appearance of the somatic chromosomes at mitotic metaphase is termed the *karyotype*. Parts of the plants with rapidly dividing cells, such as root tips, are easily prepared for cytological examination by routine techniques such as aceto-carmine squashes or by embedding in wax and sectioning (DARLINGTON and LA COUR, 1962).

A quick method for examining chromosomes in root tips is to place the fresh root tips in a solution of one part 45 per cent acetic-orcein and one part Normal HCl. This is then heated to approximately 60°C for 10 to 15 minutes and the roots can then be removed, mounted in 45 per cent acetic acid, squashed between slide and coverslip and examined under a suitable microscope.

To study pollen meiosis, flower buds should be fixed overnight in a solution of 1:3 acetic acid alcohol with a trace of ferric acetate. The anthers should then be dissected out in 45 per cent aceto-carmine, gently heated over a spirit lamp and squashed under a coverslip for examination with a microscope.

10.1 Chromosome number and morphology

The value of chromosome number as a taxonomic character is that it is usually constant within a species, although the number of exceptions is higher than was at one time believed. The numbers recorded in flowering plants range from $2n=4$ in *Haplopappus gracilis*, a composite, to $2n=265$ in the grass *Poa litorosa*, and very high numbers are found in some ferns. Chromosome number may be uniform in a genus or higher group or vary from member to member.

The basic number of chromosomes (usually referred to as x) may occur as exact multiples in different members of a group, giving what is called a polyploid series ($2x$=diploid, $3x$=triploid, $4x$=tetraploid, $5x$=pentaploid, $6x$=hexaploid, etc.) as in the Malvaceae with $x=5, 6, 7, 11$ and 13 and somatic numbers ranging from 10 to 90. A survey of the reported chromosome numbers of 17,138 Angiosperm species showed that the average haploid number of chromosomes per species for the following groups were:

Temperate and arctic Dicotyledons $n=15{\cdot}26$
Tropical Dicotyledons $n=16{\cdot}07$
All Dicotyledons $n=15{\cdot}36$
Monocotyledons $n=17{\cdot}46$
All Angiosperms $n=15{\cdot}99$ (data from GRANT, 1963)

In herbaceous Dicotyledons the basic haploid chromosome numbers show modes of $n=7$, 8 and 9, while in the woody Dicotyledons sampled the frequency distribution is bimodal with one peak at $n=8$ and 9 and the other at $n=11$–14. There are reasons for believing that the original chromosome number in the Angiosperms was in the range 7–9, and if this assumption is correct then any chromosome from about $n=14$ upwards is of polyploid origin.

Increase in chromosome number may be attained by means other than polyploidy, for example by stepwise changes due to various mechanisms. Likewise there may be progressive decreases in the basic number or in a polyploid number. In such cases where the numbers found within a group show no simple numerical relationship to each other the term *aneuploidy* is used. From this it follows that $n=14$ may not always be a polyploid number and some of the lower numbers $n=10$ to $n=13$ may be polyploids derived from reduced basic numbers of 5, 6 and 7. But even allowing for this variation, if we accept species with $n=14$ and above as polyploids, then 43 per cent of the 12,000-odd species of Dicotyledons sampled and 58 per cent of the 5,000-odd species of Monocotyledons are polyploids, giving a figure of 47 per cent for the Angiosperms as a whole. Other methods of calculation have given similar results and it is generally accepted that from a third to a half of present-day Angiosperms have had a polyploid origin. This has important bearings on considerations of evolution and speciation (p. 52).

So far chromosome counts have been made for fewer than 10 per cent of flowering plant species, but not even all of these can be relied upon as many were obtained in the early days of cytology and are probably erroneous through faulty techniques or for other reasons such as misidentification of the species. Furthermore very many of the numbers recorded have been counted only once and even then from a single plant raised in cultivation. Errors of identification are common, especially if the name given to the species in question by a botanic garden, or even a taxonomist, from which the seed is obtained, is accepted without further checking. It is important that voucher specimens should be kept for all plants whose chromosomes have been counted; in all cases the source of identification should be given since this may be as important as the identity of the person who counted the chromosomes.

Apart from their value in allowing the identity of material counted to be checked, should any doubt arise, voucher specimens may be valuable in allowing taxonomists to find out if any morphological differences are associated with cytological features as in the diploid and tetraploid species pair *Petrorhagia prolifera* and *P. nanteulii* which are clearly separable by their seed coats.

Very little intensive sampling of the cytological variation found in populations has yet been undertaken, but the evidence from those species that have been adequately sampled from various parts of their area suggests that there may be considerable lack of uniformity in chromosome numbers. This is particularly true in species which show sexual abnormalities such as apomixis, as in *Poa alpina* which has a range of chromosome numbers from 14 to 56, the frequency of the different numbers varying in different parts of the species' range. Another outstanding example is the *Cardamine pratensis* complex in which numbers of $2n = 16$, 24, 28, 30, 32, 38, 40, 42–55, 56, 60, 64, 73, 56–80, 84, 88, 90, and c. 96 have been found, although not always clearly associated with morphological characters, geographical distribution or ecological features. Many less extreme examples have been found, the commonest being where polyploid numbers occur in species previously thought to be uniform cytologically.

Another frequent occurrence is the discovery that situations where we thought we were dealing with a simple diploid-tetraploid pair of species have turned out to be only fragments of intricate polyploid complexes embracing several diploids, tetraploids and even higher levels of polyploid. These polyploid complexes are discussed later (p. 53).

A great deal of work has been done recently on the occurrence of accessory or B-chromosomes in plants. These chromosomes are normally small and heterochromatic and do not pair with the normal A-chromosomes at meiosis. They occur in variable numbers in some plants in a population but may be absent in other members of the population without apparent ill effects. Amongst the effects attributed to B-chromosomes are reduction in

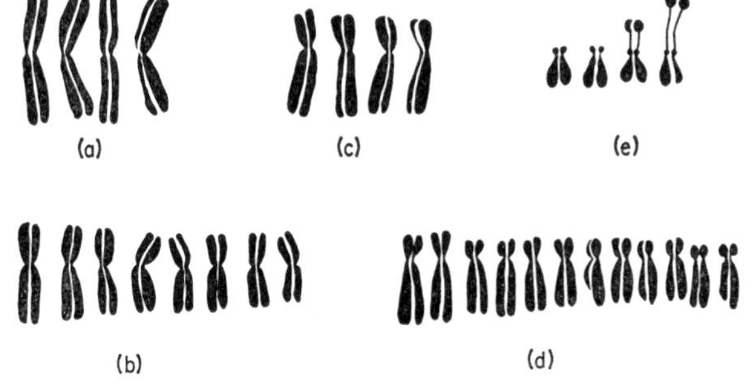

Fig. 10–1 The karyotype of *Ranunculus omiophyllus* ($2n = 32$), showing (**a**) 4 long, median centromere; (**b**) 8 medium, median centromere; (**c**) 4 long, submedian centromere; (**d**) 12 medium, submedian centromere; and (**e**) 4 short, subterminal centromere (1 pair bearing satellites). (After C. D. K. COOK, 1966, unpublished.)

fertility, increase in vigour and increase in variability. Some correlation has been shown between different frequencies of B-chromosomes and certain ecological or climatic factors such as humidity, mineral content of soil, etc.

It is evident that we still have a great deal to learn about the basic facts of chromosome number distribution in the Angiosperms and many millions of additional counts will have to be made to bring our knowledge up to a satisfactory level.

Chromosomes vary not only in number but in size, both absolute and relative, shape and volume, and in the amount and distribution of heterochromatin. These features may be observed in karyotypes in certain groups where the individual chromosomes are large enough to see in detail. In some cases it is possible to identify each individual chromosome of a karyotype by its relative size, position of the centromere, presence of satellites, etc. (see Fig. 10–1). Evolutionary trends in such features have been worked out in some groups although it is difficult at times to decide upon the direction of these trends. (For examples see DARLINGTON, 1956.)

10.2 Behaviour at meiosis

Their appearance and the way chromosomes behave at meiosis is frequently used as a means of indicating the relationship of populations and species through the kind and degree of pairing that occurs between the different parental sets. This may show whether hybridization has occurred, show structural differences, explain the causes of sterility, and suggest the derivation of one species from another. The degree of homology between the chromosomes in hybrids is used as an estimate of the degree of relationship of the parental species, irrespective of the extent of the morphological differences between them. Not enough is known, however, about the factors controlling chromosome pairing; it has been demonstrated recently that pairing in wheat, *Triticum*, can be under simple genetic control.

10.3 Polyploidy

Polyploidy, the occurrence of multiple chromosome sets in an organism, is widespread in Angiosperms as we noted above. It plays an important role as an evolutionary mechanism and at the same time it often presents problems for the taxonomist.

Most polyploids are formed as a result of the doubling of chromosomes of hybrids formed between separate species, or at least between different races of the same species. An immediate consequence of this mechanism is that the polyploid is normally unable to form fertile hybrids with the diploids due to the inability of many of the chromosomes to pair. Thus a diploid-tetraploid pair is separated by the formation of sterile triploids.

Although polyploidy is regarded as a mechanism of abrupt species formation, it is the success of the polyploids that is important, not just the doubling of the chromosomes; only if they are able to find a suitable ecological niche and establish themselves as differentiated populations will they be regarded as separate species from their parents. Initially there may be little morphological difference between the diploids and polyploids and the problem of the taxonomist is to find ways to recognize them. Sometimes polyploid pairs such as *Nasturtium officinale* ($2n=32$) and *N. microphyllum* ($2n=64$), *Cardamine hirsuta* ($2n=16$) and *C. flexuosa* ($2n=32$) can be easily separated. Often there are statistical differences in features such as pollen diameter, epidermal cell size, seed size, etc., but in many cases these differences break down when a wide range of population samples is examined from different geographical areas. The difficulties experienced in separating polyploid pairs or sequences are reflected in their treatment as subspecies or varieties of a single species, as in the case of *Ranunculus ficaria* which contains diploid, triploid, tetraploid, pentaploid and hexaploid races in Europe. Their geographical distributions overlap to some extent and in Britain only diploid and tetraploid races are common, separable by the presence of bulbil-like tubercles in the leaf axils of the latter and by a number of less reliable features such as mean petal width and number, stamen number and carpel number. As frequently happens it is easy to separate diploid and tetraploid populations in any one locality, but the features by which they may be recognized tend to vary from one locality to another.

When it has been confirmed that diploid and tetraploid populations occur in a region, these may be sampled by taking a random hundred flowers of each and scoring them for (1) number of stamens, (2) number of carpels, (3) number of petals and (4) whether the petals are broad and overlapping, intermediate, or narrow and not overlapping, as well as (5) for the presence or absence of tubercles. The results for (1), (2) and (3) can be plotted in the form of frequency histograms which are basically graphs in which the vertical axis shows the number of occurrences and the horizontal axis shows the classes (each number forming a class). The other data (4) and (5) can be tabulated and the two sets compared to give a picture of the variation in the populations.

Polyploidy is often a means by which sterile hybrids between distinct species overcome their sterility by chromosome doubling. Amongst species which have originated in this way are the *Poa annua* ($2n=28$) derived from *P. supina* ($2n=14$) and *P. infirma* ($2n=14$) and the saltmarsh grass *Spartina townsendii* ($2n=120, 2, 4$) from the European *S. maritima* ($2n=60$) and the N. American *S. alterniflora* ($2n=62$) which was introduced into Britain by accident in the nineteenth century. Many of our crop plants such as wheat, cotton, brassicas and potatoes are of polyploid origin.

Many polyploid situations have turned out to be very complex upon further investigation. A common type, referred to as a *polyploid pillar complex* occurs where a number of closely related diploid species or races support a tetraploid (and even hexaploid) superstructure.

These pillar complexes present great problems to the taxonomist because of the presence of the same or similar genomes at the different levels ($2x$, $4x$, etc.) and because of the frequent hybridization that occurs between the members at the various polyploid levels and even between levels in some cases. As a result we find that the whole structure is interconnected in terms of gene flow, and morphologically similar forms occur at the different levels because of the common genomes and the hybrids between them. Frequently it is found impossible to distinguish the polyploid structure as a whole, or individual members of it, from the diploid species on which it is based. Even when the diploid members are intersterile, their polyploid representatives often hybridize and there may develop a cyclical process consisting of differentiation at the diploid level followed by polyploidy and hybridization at the polyploid level where, as a result of selection from the array of variation, further differentiation leading to secondary species formation will take place, and the whole process can be repeated. Such differentiation-hybridization cycles have been described in many polyploid complexes including ferns such as *Dryopteris* where it appears that the polyploid members are more clearly differentiated from each other and from the diploids than in similar Angiosperm situations. It may be that the ferns being an older group, there has been more time for clear-cut secondary species formation to take place; on the other hand it is possible that further investigation will reveal intermediate stages in the process.

An excellent example of a polyploid complex is found in the water crow-foots, *Ranunculus* subgenus *Batrachium*. These plants which are all aquatic or marsh plants have a basic chromosome number of 8 and somatic counts of 16, 24, 32, 40 and 48 have been found in material collected in the wild. Polyploidy is found in some species such as *R. omiophyllus* which has $2n=16$ and 32 and *R. peltatus* with $2n=16$, 32 and 48, while in other species only a single number has been found: *R. ololeucos* $2n=16$, *R. baudotii* $2n=32$ and *R. tripartitus* $2n=48$. Artificial hybrids have been made between plants of all these species, and it has been found that crosses between the diploids are inviable while crosses between tetraploid or hexaploid plants are usually viable and often give rise to fertile hybrids. It is possible to cross diploid and tetraploid plants but the hybrids are sterile; crosses between the hexaploids and diploids or tetraploids are usually viable, giving fertile hybrids. Under experimental conditions it is therefore found that some gene exchange can take place between any one species and any other, although it is unlikely that this will happen in nature.

From the point of view of potentially common gene pools which is the basis of many 'biological species' definitions (p. 4) the whole complex has to be regarded as a single species. Using morphological criteria polyploids and diploids will be included in the same species even though they are separated from each other by a sterility barrier; conversely interfertile populations would be separated as different species on morphological grounds.

Modes of species formation

The groups of plants which taxonomists regard as species arise by a complex series of processes. We have seen how incipient species may be formed by polyploidy, followed by a gradual process of further differentiation under suitable circumstances. Incipient species may also arise abruptly through the acquisition, by an individual, of reproductive isolation caused by gene mutations leading to chromosomal rearrangements. This mechanism is possible in self-compatible plants and the chromosomal rearrangement, consisting of an alteration in the linear order of the genes, has the effect of isolating the new individuals from the parents since the hybrids formed between them would be sterile due to the inability of the chromosomes to pair at meiosis. The other principal method is wholly gradual and consists of the step-by-step accumulation of small differences caused by mutation, recombination (including hybridization) and selection, often accompanied by the slow build-up of isolating mechanisms.

Geographical isolation of populations is normally a necessary or even essential condition for the development of isolating mechanisms and thereby species formation. Small-scale adaptive variation of populations as in the formation of ecotypes has already been mentioned; larger-scale differentiation of populations is a common phenomenon in species of wide distribution (as in *Digitalis purpurea*, p. 6) and may be recognized by the description of subspecies.

Those populations that have not reached the level of differentiation or degree of reproductive isolation for them to be treated as species are referred to as *races*. Various degrees of reproductive isolation may be found between the different races of a species, and the wider separated geographically the less fertile the hybrids between them may be: there may even be complete sterility between some populations of a species.

Isolating mechanisms

There are many different isolating mechanisms which prevent or reduce gene flow between populations; they may occur in different combinations and operate at different intensities. A distinction is drawn between *spatial* or *environmental isolation* which prevents or at least reduces gene exchange between populations because of distance or the limited availability of suitable habitats, and *reproductive isolation* which depends on some aspect of the plants themselves.

Populations or species which are separated geographically are called *allopatric* as opposed to those which grow in the same areas when they are called *sympatric* (when they are in contact at their margins the term *semipatric* may be used). Isolating mechanisms may not develop in allopatric species and the hybrids formed between the species are highly fertile, while related sympatric species in the same genus may have strong incompatibility barriers and the hybrids between them are sterile.

Reproductive isolation may be *external* due to mechanisms operating outside the plant and effective before fertilization: e.g. *mechanical* devices such as lock-and-key pollination mechanisms and other structural contrivances of the flower as in many orchids; *ethological* factors where cross pollination is possible but does not occur due to the constancy of the pollinator to a particular type of flower, again as in many orchids; or *seasonal* where the flowering time of two species is sufficiently different to cut down cross pollination (flowering at different times of day, as in some *Agrostis* species, may also be effective). *Internal* mechanisms which operate inside the plants from pollination onwards cover incompatibility, hybrid inviability, hybrid sterility and hybrid breakdown (see STEBBINS, 1950, for detailed discussion).

10.4 Isolation and species definitions

Although isolating mechanisms are of great evolutionary importance since they allow populations to build up differences and evolve as independent units, the groups of plants which we recognize as species are in practice defined largely by morphological criteria. In very many cases there is a high degree of correspondence between morphological differentiation and inability to interbreed, but the two are not necessarily correlated as we have already seen. In allopatric species, for example, isolating mechanisms may not be developed and indeed, from a practical point of view, they are not necessary since the populations are prevented from hybridizing by distance. We also saw that various degrees of fertility may exist between members of different populations of the same species. In the case of polyploid situations, morphologically delimited species may embrace diploid and polyploid members which are separated by sterility barriers. Finally, in predominantly inbreeding groups, each member of a population is prevented from breeding with the others.

10.5 Hybridization

Since isolation between populations is seldom 100 per cent effective, hybridization between species is a common occurrence in nature. Hybridization in fact ranges from the occasional crossing of individuals of sympatric species producing inviable or sterile hybrids to complete breakdown of isolation between parts of species leading to the formation of hybrid swarms. Intermediate situations where there is a certain amount of interbreeding through the formation of fertile hybrids, some of which backcross to one or both parental species, are known as introgression. The hybrids formed are often incapable of producing viable seed but their pollen may be viable and able to fertilize members of the parental species. Or the hybrids may be fully fertile but unable to grow in the habitat of either parent and, if no intermediate habitats are available, hybrids in nature will be rare. This explains how species remain distinct in nature in the face of hybridization: since the hybrids are not so well adapted as their parents to their habitats, they are weeded out by natural selection. In disturbed habitats formed by deforestation, clearing, cultivation, etc., some of the hybrids formed may have a combination of characters better adapted than their parents to the new mixture of habitats produced. This is sometimes referred to as 'hybridization of the habitat' following ANDERSON (1949) who pioneered studies in introgressive hybridization.

Because of backcrossing it sometimes happens that hybrids between species resemble one or other of the parents so closely that they are not recognized as being in fact hybrids. This occurs in the *purpurea* group of the genus *Digitalis* where hybrids between *D. purpurea* and *D. thapsi* were thought to be relatively rare in nature. Artificial hybrids between them are very similar in habit and general appearance to *D. thapsi* and many plants identified as the latter have turned out to be hybrids. Detailed study of the morphological features of the hybrids has shown that they can be easily separated by submicroscopic features such as the nature, distribution and number of cells in the hairs of the sepal margins. The parental species have characteristic sepal hair profiles and that of the hybrids is intermediate (Fig. 10–2).

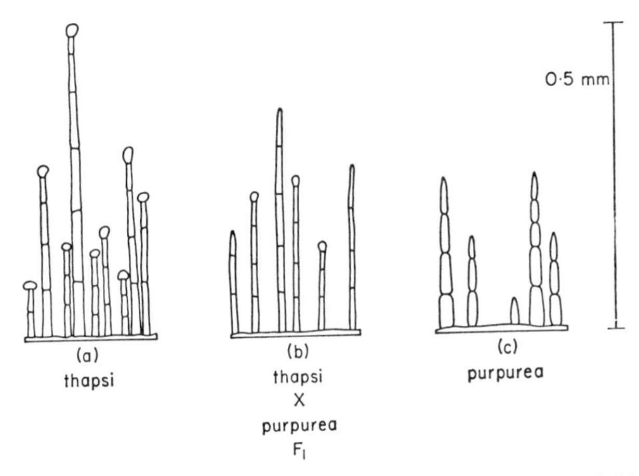

Fig. 10–2 Sepal hairs of (**a**) *Digitalis thapsi*, (**c**) *D. purpurea*, and (**b**) the F_1 hybrid between them.

Other methods for the recognition of hybrids include chromatography, mentioned on p. 36, and various graphical techniques devised by ANDERSON (1949).

One of these, the *hybrid index*, can be applied to pairs of species such as *Geum urbanum* and *G. rivale* which frequently form hybrid swarms when they occupy the same habitat. The two species may be distinguished by several morphological features of which the following six can be selected with ease:

1. Total number of leaflets of the basal leaf
2. Length of bract subtending the inflorescence
3. Length of the stipule of the bract
4. Longest sepal
5. Sepal colour
6. Longest petal

Each character can be divided into classes as shown in Table 3 so that *G. urbanum* obtains the lowest score and *G. rivale* the highest score when converted into a hybrid index. This is only an example and in practice different values may be found from those given here.

Table 3

	Score				
Character	0	1	2	3	4
1.	up to 5	6–8	9–11	12–14	15 or more
2.	60 mm or more	49–60 mm	37–48 mm	25–36 mm	24 mm or more
3.	29–35 mm	22–28 mm	15–21 mm	8–14 mm	0–7 mm
4.	4.5–6.5 mm	7–9 mm	9.5–11.5 mm	12–14 mm	14.5–16.5 mm
5.	green		intermediate		red
6.	0–5 mm	6–7 mm	8–9 mm	10–11 mm	12–13 mm
	G. urbanum				*G. rivale*

Thus 'pure' *G. urbanum* will have a hybrid index of 0 and 'pure' *G. rivale* an index of 28. If the index values for plants sampled from the hybridizing populations are plotted in the form of a frequency distribution histogram, a picture of the approximate extent of hybridization will be obtained.

A notable feature of modern systematics is the way in which data from genetics, cytology, biometrics, chemistry and ecology have contributed to our understanding of the mechanisms by which organisms, populations and species evolve and change. The major outlines are known but an enormous amount of detail has still to be filled in.

Further Reading

CORNER, E. J. H. (1964). *The Life of Plants*. Wiedenfeld & Nicolson, London.
DAVIS, P. H. and HEYWOOD, V. H. (1965). *Principles of Angiosperm Taxonomy*. Oliver & Boyd, Edinburgh & London.
EHRLICH, P. R. and HOLM, P. W. (1963). *The Process of Evolution*. McGraw–Hill Book Co. Inc., New York & Maidenhead.
GRANT, V. (1963). *The Origin of Adaptations*. Columbia University Press, New York & London.
HESLOP-HARRISON, J. (1953). *New Concepts in Flowering Plant Taxonomy*. Heinemann, London.
SOLBRIG, O. T. (1966). *Evolution and Systematics*. MacMillan, New York & London.
STEBBINS, G. L. (1950). *Variation and Evolution in Plants*. Columbia University Press, New York; Oxford University Press, London.

References

ALSTON, R. E. and TURNER, B. L. (1963). *Biochemical Systematics*. Prentice-Hall Inc., New Jersey.
ANDERSON, E. (1949). *Introgressive Hybridization*. Wiley, New York; Chapman & Hall, London.
BAILEY, I. W. (1954). *Contributions to Plant Anatomy*. Chronica Botanica, Waltham, Mass.
CARLQUIST, S. (1961). *Comparative Plant Anatomy*. Holt, Rinehart & Winston, New York.
DARLINGTON, C. D. (1956). *Chromosome Botany*. Allen & Unwin, Ltd., London.
DARLINGTON, C. D. and LA COUR, L. F. (1962). *The Handling of Chromosomes*. Allen & Unwin, Ltd., London.
DAUBENMIRE, R. F. (1959). *Plants and Environment. A Textbook of Plant Autecology*. Wiley, New York; Chapman & Hall, London.
DUPRAW, E. J. (1965). Non-Linnean taxonomy and the systematics of honey-bees. *Systematic Zoology*, **14**, 1–24.
EHRLICH, P. R. and RAVEN, P. H. (1965). Butterflies and Plants: A study in coevolution. *Evolution*, **18**, No. 4, 586–608.
ERDTMAN, G. (1963). Palynology. In *Advances in Botanical Research* edited by R. D. PRESTON, **1**, 149–208.
GELL, P. G. H., HAWKES, J. G. and WRIGHT, S. T. C. (1959). The application of immunological methods to the taxonomy of species within the genus *Solanum*. *Proc. Roy. Soc. Ser. B*, **151**, 364–383.
GOULD, S. W. (1962). Family names of the Plant Kingdom. *International Plant Index*. 1, New Haven & New York.
GRANT, V. (1958). The regulation of recombination in plants. *Cold Spring Harbor Symp. Quant. Biol.*, **23**, 337–363.

HEGNAUER, R. (1962–1966). *Chemotaxonomie der Pflanzen*, vols. I–IV. Birkhauser, Basel.

HEYWOOD, V. H. and MCNEILL, J., eds. (1964). *Phenetic and Phylogenetic Classification. Syst. Ass. Publ.* **6**.

HOYER, B. H., MCCARTHY, B. J. and BOLTON, E. T. (1964). A molecular approach in the systematics of higher organisms. *Science*, **144**, 959–967.

LEONE, C. A. (1964). *Taxonomic Biochemistry and Serology*. Ronald Press, New York.

MAHESHWARI, P., ed. (1963). *Recent advances in the embryology of Angiosperms*. International Society of Plant Morphologists. University of Delhi, Delhi.

MEEUSE, A. D. J. (1965). *Angiosperms—Past & Present. Phylogenetic Botany and Interpretative Floral Morphology of the Flowering Plants*. Institute for the advancement of Science and Culture, New Delhi.

MELVILLE, R. (1962). A new theory of the Angiosperm flower, I. The Gynoecium. *Kew Bull.*, **16**, 1–50.

PERRING, F. H. and WALTERS, S. M., eds. (1962). *Atlas of the British Flora*. Botanical Society of the British Isles, Nelson, London & Edinburgh.

SNEATH, P. H. A. (1964). New approaches to bacterial taxonomy: use of computers. *Ann. Rev. Microbiol.*, **18**, 335–346.

SOKAL, R. R. and SNEATH, P. H. A. (1963). *Principles of Numerical Taxonomy*. Freeman, San Francisco & London.

SOKAL, R. R. and SNEATH, P. H. A. (1966). Efficiency in taxonomy. *Taxon*, **15**, 1–21.

SPORNE, K. R. (1956). The phylogenetic classification of the Angiosperms. *Biol. Rev.*, **31**, 1–29.

STACE, C. A. (1965). Cuticular studies as an aid to plant taxonomy. *Bull. Mus. Brit. Nat. Hist. Bot.*, **4**, No. 1.

STIX, E. (1960). Pollenmorphologische Untersuchungen an Compositen. *Grana Palynologica*, **22**, 41–114.

SWAIN, T., ed. (1963). *Chemical Plant Taxonomy*. Academic Press, New York & London.

SWAIN, T., ed. (1966). *Comparative Phytochemistry*. Academic Press, New York & London.

ZUCKERKANDL, E. (1965). The evolution of haemoglobin. *Scientific American*, **212**, 110–119.